Chemistry-Biology
Interface Series

Series Editor
Charles C. Price
Department of Chemistry
University of Pennsylvania

Bender and Brubacher	CATALYSIS AND ENZYME ACTION
Benfey	INTRODUCTION TO ORGANIC REACTION MECHANISMS
Clayton	LIGHT AND LIVING MATTER, Volume 1: The Physical Part
Clayton	LIGHT AND LIVING MATTER, Volume 2: The Biological Part
Price	GEOMETRY OF MOLECULES

CATALYSIS AND ENZYME ACTION

MYRON L. BENDER
Professor of Chemistry
Northwestern University

LEWIS J. BRUBACHER
Assistant Professor of Chemistry
University of Waterloo

McGRAW–HILL BOOK COMPANY
New York *St. Louis* *San Francisco* *Düsseldorf* *Johannesburg* *Kuala Lumpur* *London* *Mexico* *Montreal* *New Delhi* *Panama* *Rio de Janeiro* *Singapore* *Sydney* *Toronto*

This book was set in Press Roman by Scripta Technica, Inc.
The editors were James R. Young, Jr., and J. W. Maisel;
the designer was Barbara Ellwood;
and the production supervisor was Joe Campanella.
The drawings were done by Oxford Illustrators Limited.
The printer and binder was The Murray Printing Company.

LIBRARY OF CONGRESS CATALOGING IN PUBLICATION DATA

Bender, Myron L 1924–
Catalysis and enzyme action.

(Chemistry–biology interface series)
Includes bibliographies.
1. Catalysis. 2. Enzymes. I. Brubacher, Lewis J., 1937– joint author. II. Title.
QD505.B46 541'.395 73–6607
ISBN 0–07–004450–3
ISBN 0–07–004451–1 (pbk.)

CATALYSIS AND
ENZYME ACTION

2 3 4 5 6 7 8 9 0 M U M U 7 9 8 7 6 5 4 3

to our wives

MURIEL S. BENDER
and
LOIS M. BRUBACHER

CONTENTS

The Chemistry-Biology Interface Series

Several years ago, a few dozen biologists, chemists, physicists, and other scientists spent several days on the campus of the University of Washington under the joint sponsorship of the Commission on Undergraduate Education in Biology, the Advisory Council on College Chemistry, and the Commission on College Physics. The purpose was to study ways to improve teaching in areas of mutual concern to two or more of the disciplines involved. The group considering the area between chemistry and biology agreed that a series of paperback books, prepared for elementary college level students in either biology or chemistry could serve a useful purpose toward this end.

Prepared by authorities in their fields, these books could, for the chemists, indicate the biologically significant reactions useful to illustrate chemical principals and, for the biologist, summarize up-to-date information on molecular phenomena of significance to a modern understanding of biological systems.

To implement this proposal, CUEBS and AC_3 appointed an editorial committee of:

to organize the undertaking.

As of this writing, the following volumes have been published:
O. T. Benfey, "INTRODUCTION TO ORGANIC REACTION MECHANISMS"
Roderick K. Clayton, "LIGHT AND LIVING MATTER," Vols. I and II
Charles C. Price, "GEOMETRY OF MOLECULES"

The following volumes are planned:
Melvin Calvin, "CHEMICAL EVOLUTION"
Paul M. Doty, "MACROMOLECULES"
David E. Greene, "SURFACE, FILMS AND MEMBRANES"

It is our hope that the material in these volumes will prove of sufficient interest to teachers and students in elementary college chemistry and biology courses that much of it will ultimately be incorporated in regular textbooks.

CHARLES C. PRICE
Philadelphia, Pennsylvania

PREFACE

It is the aim of this book to introduce college students in chemistry and biology to the concepts of catalysis, that is, how a reaction rate is accelerated. The catalysts vary from the very simple proton to the very complex enzymes and heterogeneous catalysts, both containing thousands of atoms.

The enzymes are biocatalysts; that is, they catalyze most of the body's functions. Heterogeneous catalysts are usually solid materials that are commercially very important. These catalysts are responsible for about 100 billion dollars worth of products annually in our economy.

In between these two extremes are catalysis by bases, such as hydroxide ions, general (undissociated) acids, general bases (usually organic compounds), metal ions, organic nucleophiles and electrophiles, as well as more complicated forms of catalysis, such as multiple catalysis, intramolecular catalysis, and intracomplex catalysis. Metal ions can act as catalysts by serving as superacids (protons of magnified charge in neutral solution), by acting as electron transfer agents, or by acting as templates on which reactions occur. Catalysis by nucleophiles and electrophiles includes a discussion of two important vitamins, thiamine pyrophosphate (B_1) and pyridoxal phosphate (B_6). Multiple catalysis, as the name implies, propounds that if a little is good, a lot is better. Intramolecular catalysis has demonstrated some important models of complex systems in which the catalyst is tied down in the same molecule as the bond to be broken. Intracomplex catalysis is closer to enzyme catalysis because in this instance, a noncovalent complexing is a necessary prerequisite for the subsequent catalytic steps.

Enzymic catalysis, including the structure, kinetics, and inhibition of enzymes, is discussed. Like intracomplex catalysis, an enzyme forms a noncovalent complex with the molecule it is acting on, called the substrate. Enzymes are usually protein in nature, being polymeric materials composed of amino acid monomers. But sometimes small cofactors, either organic or inorganic, are needed for enzyme action. These substances, called coenzymes, can range from complex metal ions such as vitamin B_{12} to the vitamins mentioned above (B_1 and B_6). Catalysis by two enzymes, chymotrypsin and lysozyme, is described in some detail, including how the fit between the enzyme and the substrate is important. The factors by which enzymes achieve their catalytic efficiency and specificity are also emphasized.

Other books in this series emphasize the geometry of molecules, organic reaction mechanisms, macromolecules, etc. This book attempts to build on the other members of this series.

MYRON L. BENDER
LEWIS J. BRUBACHER

ONE
INTRODUCTION

In order to understand how catalysts function, we must first understand what happens to the atoms in a molecule during a typical chemical reaction.

1-1 THE ANATOMY OF A CHEMICAL REACTION

As an example of a typical chemical reaction we shall consider the hydrolysis of methyl iodide in water solution to give methyl alcohol. The equation for this reaction, as written in Eq. (1-1), tells us nothing about what happens to the individual atoms during the course of a reaction between one molecule of methyl iodide and one molecule of water.

$$\underset{\text{Methyl iodide}}{CH_3I} + \underset{\text{Water}}{H_2O} \longrightarrow \underset{\text{Methyl alcohol}}{CH_3OH} + \underset{\text{Hydrogen iodide}}{HI} \qquad (1\text{-}1)$$

Equation (1-2) illustrates what happens more clearly. In general a chemical reaction involves the breaking of old bonds and the making of new bonds, which may take place in several steps. The reaction in Eq. (1-1) is relatively simple. It occurs in a single step, as shown in Eq. (1-2). The intermediate stage in which two bonds are breaking while another is forming is called the *transition state*. We shall define this term more precisely later in this chapter.

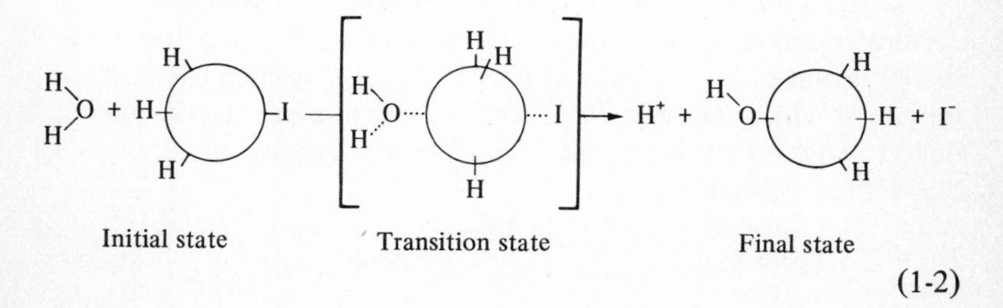

(1-2)

Let us center our attention on the methyl iodide molecule in Eq. (1-2). The carbon atom is represented by a circle, and four atoms are bonded to it in a tetrahedral arrangement, as indicated by the drawing in the initial state. When a water molecule approaches the methyl iodide molecule from its backside (relative to the iodine atom substituent) with sufficient energy, a new bond begins to form as indicated in the drawing labeled "transition state." Simultaneously, the bond between the iodine atom and the carbon atom weakens. In the transition state the carbon atom of methyl iodide has three more or less normal bonds (to the three hydrogen atoms) and two partial bonds, the bond to the iodine atom which is breaking, and the bond to the oxygen atom which is forming. Although the total system is electrically neutral, the iodine atom is beginning to acquire a partial negative charge and the hydrogen atom of the water molecule a partial positive charge. When the oxygen-carbon bond is completely formed and the carbon-iodine bond is completely broken, the final state is reached and the reaction is complete.

In one aspect Eq. (1-2) is incomplete. The H^+ ion is not a free species but is covalently bonded to a water molecule as an hydronium ion, H_3O^+. That is, the proton is transferred from the forming methyl alcohol molecule to a neighboring H_2O molecule. Such a proton transfer reaction occurs very readily and may be viewed as occurring almost simultaneously with the breaking and forming of the two bonds involving the central carbon atom in Eq. (1-2).

The way Eq. (1-2) is written establishes a convention that will be used in this book. The transition-state representation is shown in large brackets within the arrow. This is to emphasize that the transition state is not an intermediate compound, but is a structure through which the reactant(s) pass in the 10^{-13} s or so that it takes to go from the initial state to the final state.

The description in this section of the course of events during a chemical reaction is called the *mechanism* of the reaction (see O. T. Benfey in this series). A study of the *kinetics* of a reaction is a study of the rate at which the reaction occurs, i.e., the number of molecules of methyl iodide that react per second under various conditions of temperature, solvent, etc.

1-2 THE ENERGY REQUIREMENTS OF A CHEMICAL REACTION

It takes energy to break a chemical bond such as the carbon-iodine bond in Eq. (1-2). Conversely, energy is released when a bond, such as the carbon-oxygen bond in Eq. (1-2), is formed. In the initial stages of the reaction in Eq. (1-2), the energy being released by the forming bond is *not* sufficient to begin to break the carbon-iodine bond. Additional energy is required, and it comes from the kinetic energy of the two molecules. Thus, if a reaction is to occur, the water and methyl iodide molecules must collide with sufficient force to provide the additional energy. In this particular reaction, the additional energy requirement, or *energy barrier*, is rather large. Consequently, only a very small fraction of all collisions involve sufficient kinetic energy for the reaction to occur and the reaction rate is low. At 25°C it takes 108 d for half the methyl iodide molecules to react. This is the half-life of the reaction. Raising the temperature of the solution increases the number of sufficiently energetic collisions which occur per second. At 50°C the reaction rate is 35 times faster than at 25°C; the half-life of the reaction is only 3.1 d.

The energy requirements for the reaction of a single molecule according to Eq. (1-2) may be represented graphically as in Fig. 1-1. Energy is plotted along the ordinate; the abscissa, labeled "reaction coordinate," represents the extent of reaction. Thus the curve in Fig. 1-1 shows schematically the minimum energy required by a molecule of water plus a molecule of methyl iodide as they go from the initial state to the transition state and on to the final state. The point of maximum energy on this curve corresponds to the transition state. For convenience we have assigned zero energy to the initial state; since the final state is at lower energy, a small amount of energy is actually released in the course of the reaction. Our immediate interest, however, is in the energy barrier to reaction, 4.52×10^{-20} cal per molecule of methyl iodide or 27.18 kcal mol^{-1}, which is the difference between the energy of the transition state and the energy of the initial state. This energy difference is referred to as the *free energy of activation* $\Delta G^{\ddagger}$. It is one of the few aspects of Fig. 1-1 which can be quantitatively determined. Diagrams like Fig. 1-1 are only schematic representations. The extent of reaction of a single molecule (i.e., the abscissa) is not a directly measurable quantity. It does not correspond directly to the dimensions of either space or time; however there is a qualitative relation between the extent of reaction and the distance between the carbon and iodine atoms. We can say with simple certainty that for a given pair of reactants there is an initial state where the extent of reaction is 0 percent, and a final state where the extent is 100 percent, and that in going from the former to the latter, a molecule requires a certain minimum energy input $\Delta G^{\ddagger}$.

The free energy of activation is composed of two types of energy according to Eq. (1-3), where $\Delta H^{\ddagger}$ is the enthalpy of activation, T is the absolute temperature (in degrees Kelvin), and $\Delta S^{\ddagger}$ is the entropy of activation.

$$\Delta G^{\ddagger} = \Delta H^{\ddagger} - T\,\Delta S^{\ddagger} \tag{1-3}$$

Enthalpy is simply heat energy, the manifestation of molecular motion. At 25°C, $\Delta H^{\ddagger}$ is 27.37 kcal mol^{-1} for the hydrolysis of methyl iodide.

Entropy is a quantity that is difficult to visualize but may be thought of as a measure of disorder or randomness of orientation in a system. As a system becomes more disordered, its entropy increases. Conversely, as a system becomes more ordered, the entropy decreases. At 25°C, $\Delta S^{\ddagger}$ for the hydrolysis of methyl iodide is +0.62 cal deg^{-1} mol^{-1}; at 50°C it is -3.94 cal deg^{-1} mol^{-1}. A brief rationalization of

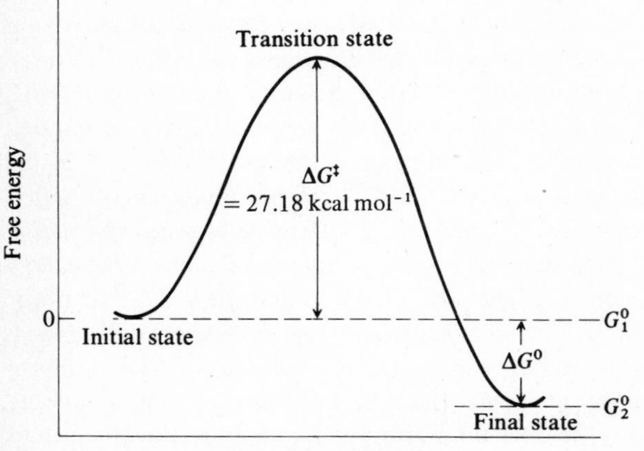

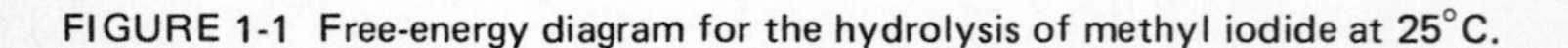
FIGURE 1-1 Free-energy diagram for the hydrolysis of methyl iodide at 25°C.

why $\Delta S^\ddagger$ decreases as temperature increases will give the reader some feeling for the concept of entropy as a measure of disorder.

The methyl iodide reactant molecule is always surrounded by a changing layer of water molecules, one of which will eventually become a reactant. This layer is the solvent shell. In the process of going from the initial state to the transition state we must ask what happens to this solvent shell as well as what happens to the reactants themselves. In other words, $\Delta S^\ddagger$ (and $\Delta H^\ddagger$, too, for that matter) has two components, $\Delta S^\ddagger_{\text{solvent shell}}$ and $\Delta S^\ddagger_{\text{reactants}}$ [Eqs. (1-4) to (1-6)].

$$\Delta S^\ddagger = \Delta S^\ddagger_{\text{solvent shell}} + \Delta S^\ddagger_{\text{reactants}} \tag{1-4}$$

$$\Delta S^\ddagger_{\text{solvent shell}} = (\text{entropy of solvent shell in transition state}) - (\text{entropy of solvent shell in initial state}) \tag{1-5}$$

$$\Delta S^\ddagger_{\text{reactants}} = (\text{entropy of reactants in transition state}) - (\text{entropy of reactants in initial state}) \tag{1-6}$$

According to one theory now in favor, the solvent shell has a certain degree of order in the initial state, perhaps similar to the ordering of water molecules in ice; in the transition state, this order breaks down somewhat (entropy increases). Therefore, $\Delta S^\ddagger_{\text{solvent shell}}$ is always positive. On the other hand, the two reactants must be lined up in just the right way in the transition state (high order, low entropy) but may have any orientation in the initial state (high disorder, high entropy). Therefore, $\Delta S^\ddagger_{\text{reactants}}$ is always negative. Since $\Delta S^\ddagger$ is slightly positive experimentally at 25°C, $\Delta S^\ddagger_{\text{solvent shell}}$ must have just a slightly greater magnitude than $\Delta S^\ddagger_{\text{reactants}}$.

Now increasing the molecular motion in the system (by raising the temperature) will increase the disorder (entropy) of the solvent shell in the initial state more than in the transition state, which has less order to start with. Thus $\Delta S^\ddagger_{\text{solvent shell}}$ will decrease in magnitude as the temperature is raised. On the other hand, $\Delta S^\ddagger_{\text{reactants}}$ should, if anything, also become more negative as the temperature is raised. Thus $\Delta S^\ddagger$ should become more negative as the temperature is increased. This is indeed what is observed. Consequently, the idea that entropy is a measure of disorder is a reasonable one.

In summary, we have described some aspects of the mechanism of the hydrolysis of methyl iodide. The course of events in bond-making and bond-breaking is illustrated by Eq. (1-2). Quantitatively 27.18 kcal of energy is required on the average to take 1 mol of methyl iodide molecules from the initial state to the transition state at 25°C. A little more than this amount of energy is released, however, in going from the transition state to the final state, so that the net result of this reaction is the release of some free energy. The rate of a reaction at a given temperature is determined by $\Delta G^{\ddagger}$. The larger $\Delta G^{\ddagger}$ is, the fewer collisions per second there are that will have sufficient energy to attain the transition state, and the slower the reaction.

1-3 THE BASIS OF CATALYSIS

In 1902 the German chemist Ostwald gave the first adequate definition of a catalyst: "a substance which alters the velocity (rate) of a chemical reaction without appearing in the end product." According to this definition a catalyst may either increase or decrease the velocity of a chemical reaction. However, in current usage, a catalyst is a substance which increases the reaction velocity; a substance which decreases the rate of a reaction is called an inhibitor. Ostwald's definition also implies that a catalyst is not consumed during the course of the reaction it catalyzes, but serves repeatedly to assist molecules to react. There are other substances which may also properly be called catalysts, but which are consumed in the reactions they catalyze. In biochemistry this is true of many coenzymes. However, these coenzymes are often restored to their original form by a subsequent reaction, so that in the larger context the coenzyme is unchanged. A typical example is the coenzyme nicotinamide adenine dinucleotide which is chemically reduced during reactions it helps to catalyze. The reduced form is oxidized back to its initial form in a subsequent reaction. In later chapters we shall meet many other coenzymes.

A catalyst increases the velocity of a reaction by increasing the number of conversions of reactant molecules to product molecules that occur each second. In order to do this the catalyst must in some way reduce $\Delta G^{\ddagger}$ for the reaction it catalyzes.

Most commonly the catalyst performs this function by providing a different pathway for the reaction, which will usually have more steps than the uncatalyzed pathway. An essential feature of the catalyzed pathway is that all $\Delta G^{\ddagger}$ values are smaller than the largest $\Delta G^{\ddagger}$ value of the uncatalyzed pathway.

To illustrate these features of a catalyst we shall consider the bromide ion as a catalyst for the hydrolysis of methyl iodide. The catalytic pathway is shown in Eqs. (1-7) and (1-8). In the first step of the reaction [Eq. (1-7)] a bromide ion reacts with a methyl iodide molecule to form a molecule of methyl bromide. The latter then reacts with water [Eq. (1-8)] to give back a bromide ion and produce the final product, methyl alcohol. The sum of Eqs. (1-7) and (1-8) is the same as Eq. (1-1); the bromide ion is not consumed in the reaction.

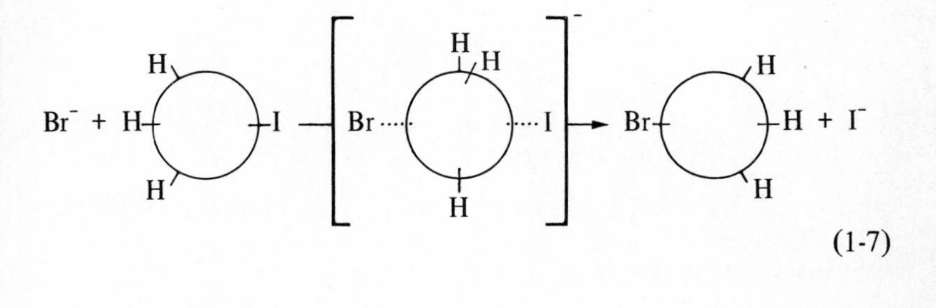

(1-7)

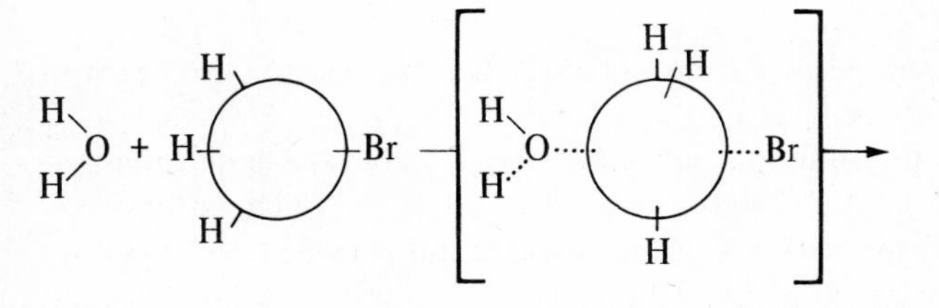

H^+ + H–O–(H)(H)H + Br^- (1-8)

$$\not{Br}^- + H_2O + CH_3I \longrightarrow H^+ + CH_3OH + I^- + \not{Br}^- \qquad (1\text{-}9)$$

Since both Eqs. (1-7) and (1-8) have lower free energies of activation than Eq. (1-2) (the uncatalyzed reaction), the rate of conversion of methyl iodide to methyl alcohol is increased by the participation of

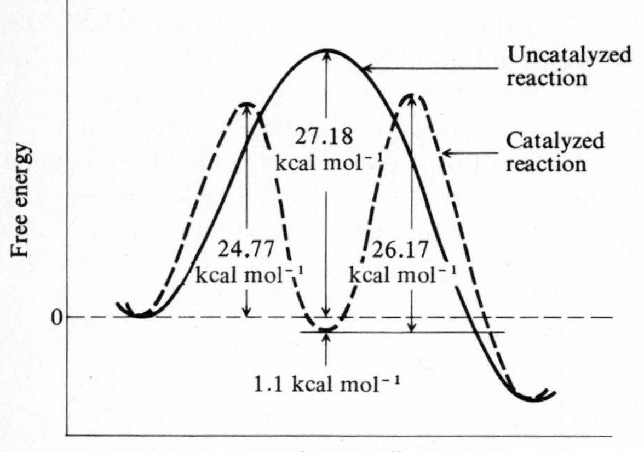

FIGURE 1-2 Free-energy diagrams for the hydrolysis of methyl iodide. Solid line: uncatalyzed; dashed line: catalyzed by bromide ion ($\Delta G^{\ddagger}$ of first step is for concentrations of 0.1 *M* bromide ion and 1×10^{-3} *M* methyl iodide).

the bromide ion. Thus the bromide ion is a catalyst for this reaction in aqueous solution.†

The transition-state diagrams for the uncatalyzed and catalyzed hydrolyses of methyl iodide are shown in Fig. 1-2. (The $\Delta G^{\ddagger}$ value for the first step of the catalyzed reaction, which depends on reactant concentrations, has been calculated for the arbitrarily chosen concentrations of 1×10^{-3} *M* methyl iodide and 0.1 *M* bromide ion.) The difference between $\Delta G^{\ddagger}$ for the uncatalyzed pathway (27.18 kcal mol^{-1}) and the largest $\Delta G^{\ddagger}$ for the catalyzed pathway (26.17 kcal mol^{-1}) is relatively small. This means that the bromide ion is not a very good catalyst. In fact the reaction via the catalyzed pathway is only about five times faster than the uncatalyzed reaction.

†This statement is correct! Most textbooks which discuss this reaction say, incorrectly, that iodide ion is a catalyst for methyl bromide hydrolysis, in the belief that methyl iodide hydrolyzes faster than methyl bromide. This error apparently arose from a misreading of the original research report [E. A. Moelwyn-Hughes, *Proceedings of the Royal Society, 164A*, 295 (1938)] which clearly states that methyl bromide hydrolyzes faster than methyl iodide in water. It is an interesting example of the perpetuation of an error through generations of textbooks. The moral is that textbook authors, ourselves included, are fallible.

Of the two steps in the catalyzed reaction of Eqs. (1-7) and (1-8), the latter has the larger free energy of activation at the specified concentrations; consequently, it is the slower step in the pathway, and it is called the *rate-determining step*.

We summarize this section by stating that *the function of a catalyst is to provide a new reaction pathway in which* (1) *the rate-determining (slowest) step has a lower free energy of activation than the rate-determining step of the uncatalyzed reaction,* and (2) *all transition-state energies in the catalyzed pathway are lower than the highest one of the uncatalyzed pathway.* (The latter condition is included for the sake of completeness even though it is not readily apparent in the example we have used.)

1-4 AN ANALOGY

A crude analogy is helpful to visualize the relationship between the potential energy of a methyl iodide molecule and its reaction pathway. A desert terrain with its humps (sand dunes) and depressions is analogous to the potential-energy possibilities available to a molecule. The average potential energy of a methyl iodide molecule in water corresponds to the height above sea level of the bottom of a depression. Passing from this hole via the path of lowest altitude to an adjacent depression corresponds to the reaction of methyl iodide with water to give methyl alcohol. A transition-state diagram is a cross section of the terrain cut along the path which the molecule takes.

Adding a catalyst (bromide ion) to the aqueous methyl iodide solution has no effect on the reactant and product potential-energy depressions or the "pass" connecting them. The presence of the catalyst is analogous to the formation of a new depression adjacent to both the product and reactant depressions to which it is connected by passes. The depth of this new depression corresponds to the average potential energy of methyl bromide plus iodide ion in aqueous solution. The passes leading to and from this depression represent the catalytic pathway. Highly energetic methyl iodide molecules can still proceed directly to products via the direct pathway, but most molecules will react via the catalytic pathway.

Other reactions of methyl iodide are conceivable; they would correspond to the reactant moving from its initial depression to another depression (such as one corresponding to the conceivable products

$CH_3CH_3 + I_2$). However, since no product other than methyl alcohol is observed, passes leading to such depressions must be much higher than that leading to methyl alcohol.

1-5 MATHEMATICAL DESCRIPTION OF REACTION RATES

The rate at which the reaction of Eq. (1-1) proceeds is proportional to the concentrations of the two reactants, methyl iodide and water. This statement is repeated mathematically in Eq. (1-10) where k' is a proportionality constant called the specific rate constant, and $[H_2O]$ symbolizes the molar concentration of water, etc. For those students who have not studied calculus, the symbol $d[CH_3I]/dt$ may be read as "the rate of change of methyl iodide concentration with time." The minus sign is used to take account of the fact that methyl iodide disappears during the course of the reaction.

$$\text{Rate} = -\frac{d[CH_3I]}{dt} = k'[H_2O][CH_3I] \qquad (1\text{-}10)$$

The hydrolysis of methyl iodide is a second-order reaction because the rate is dependent on the first power of the concentration of both reactants. In practice, however, the methyl iodide concentration would be about 1×10^{-3} *M*, whereas the water concentration is about 55 *M*. Since only 1×10^{-3} mol of water would be consumed in every liter, the concentration of water changes negligibly during the course of the reaction. The reaction is said to be pseudo first order because the rate of reaction will appear to depend only on the concentration of methyl iodide. Therefore,

$$-\frac{d[CH_3I]}{dt} = k[CH_3I] \qquad (1\text{-}11)$$

where

$$k = k'[H_2O] \qquad (1\text{-}12)$$

1-6 EQUILIBRIUM CONSTANT

Although the hydrolysis of methyl iodide goes essentially to completion in aqueous solution, the reverse reaction is believed to occur to a very slight extent. This is illustrated in Eq. (1-13), where the forward and reverse reactions are characterized by rate constants k_1 and k_{-1}, respectively.

$$H_2O + CH_3I \underset{k_{-1}}{\overset{k_1}{\rightleftharpoons}} CH_3OH + HI \tag{1-13}$$

Thus we may define an equilibrium constant for this reaction in the usual way.

$$K = \frac{k_1}{k_{-1}} = \frac{[CH_3OH]\,[HI]}{[H_2O]\,[CH_3I]} \tag{1-14}$$

The products of the hydrolysis of methyl iodide, namely, methyl alcohol and hydrogen iodide, have a lower total free energy than the reactants; this is apparent in Fig. 1-1. Thus the energy barrier is greater for the reverse reaction than for the forward reaction in Eq. (1-13). Therefore k_{-1} is less than k_1, and the equilibrium constant is much larger than unity. It is clear that these statements are all consistent with the statement at the beginning of this section that the hydrolysis of methyl iodide goes essentially to completion. Furthermore, one can grasp intuitively that the equilibrium constant is related to the difference in free energy between the initial state and the final state. The actual mathematical relationship is given in Eq. (1-15), where $\Delta G°$ is the standard free energy of the final state minus the standard free energy of the initial state ($G_2^\circ - G_1^\circ$ in Fig. 1-1), T is the temperature in degrees Kelvin, and R is the gas constant, namely, 1.98 cal deg^{-1} mol^{-1}. The numerical constant comes from the logarithm to the base 10 of the natural base, e [Eq. (1-16)].

$$\log K = -\frac{\Delta G°}{2.303RT} \tag{1-15}$$

$$\log e = \frac{1}{2.303} \tag{1-16}$$

The presence of a small amount of a catalyst affects only the height of the energy barrier and the pathway of the reaction: it ideally does not change the free energies of the initial or final states and therefore does not affect ΔG°. Consequently, a catalyst has no effect on an equilibrium constant. Thus a catalyst speeds up the reverse reaction in Eq. (1-13) to exactly the same extent as it speeds up the forward reaction.

1-7 INFLUENCE OF TEMPERATURE ON REACTION RATES

The rates of almost all chemical reactions increase as the temperature is increased; i.e., the rate constant increases with temperature. In terms of the transition-state theory, the approach we have used thus far in this chapter, the mathematical relationship between k and T is given by Eq. (1-17), where k_B, or R/N, is the Boltzmann constant (1.38×10^{-16} erg deg^{-1}), h is the Planck constant (6.62×10^{-27} erg$\cdot$s) and e is the base of natural logarithms (2.718 ...). Thus if k is known at any temperature, $\Delta G^{\ddagger}$ can be readily calculated.

$$k = \frac{k_B T}{h} e^{-(\Delta G^{\ddagger}/RT)} \qquad (1\text{-}17)$$

Another important equation relating k and T with which the reader may already be familiar is Eq. (1-18), which was proposed by Arrhenius in the late nineteenth century; E_a is the Arrhenius activation energy and A is simply called a preexponential factor. Both E_a and A are approximately independent of temperature. They have been determined for many reactions.

$$k = A e^{-(E_a/RT)} \qquad (1\text{-}18)$$

The Arrhenius equation is an empirical equation which is not derived from any theory but has been found to apply to most simple chemical reactions. On the other hand, Eq. (1-17) is derived from transition-state theory. Without presenting any proof, we shall simply note that E_a and $\Delta H^{\ddagger}$ are related by Eq. (1-19).

$$\Delta H^{\ddagger} = E_a - RT \qquad (1\text{-}19)$$

The reader will then find it an interesting mathematical exercise to show, using Eqs. (1-3) and (1-16) to (1-19), that Eq. (1-20) is the relation between A and $\Delta S^{\ddagger}$. Thus A is related to the entropy part of the free energy of activation.

$$\Delta S^{\ddagger} = 2.303R \left[\log A - \log \left(\frac{k_B T}{h} \right) \right] - R \tag{1-20}$$

SUGGESTED READINGS

Klotz, I. M.: "Energy Changes in Biochemical Reactions," Academic Press, Inc., New York, 1967. This little book introduces enthalpy, entropy, and free energy in a clear way at an elementary level and then applies these concepts to various thermodynamic calculations. Many sample calculations are given.

King, E. L.: "How Chemical Reactions Occur," W. A. Benjamin, Inc., New York, 1964. An elementary-level discussion of reaction rates and mechanisms which is in many ways complementary to what is discussed in this and later chapters.

Leisten, J. A.: "Homogeneous Catalysis. A Reexamination of Definitions," *Journal of Chemical Education,* vol. 41, p. 23, 1964.

TWO
ENZYMES AS CATALYSTS

Some catalysts are simple, others are very complex in the sense that they incorporate several different types of catalytic components. In the next chapter, we shall consider several types of simple catalysis; later in that chapter we shall see that the complex, biologically important catalysts, enzymes, may be viewed as a collection of several simple types of catalysts collected into one molecule. Enzymes do not behave magically, but according to established chemical principles. To prepare ourselves for the study of enzymes as catalysts we shall examine the structure of enzymes, some types of reactions they catalyze, and some aspects of their catalytic action.

2-1 ENZYME STRUCTURE

The simplest enzymes are proteins of molecular weight ranging from about 12,000 to 40,000. Proteins are composed of small building blocks, or residues, known as amino acids, which range in molecular weight from 75 to 200. Consequently, most simple enzymes are built up of 100 to 400 amino acid residues. The 20 amino acids found in proteins are listed in Table 2-1. In a protein the amino acids are joined together via amide, or *peptide*, bonds as illustrated in Fig. 2-1*b*. Note that in the synthesis of proteins a molecule of water is eliminated for each peptide bond that is formed, or conversely, when a protein is chemically broken down into its constituent amino acids (i.e., hydrolyzed), one molecule of water is used up for each peptide bond that is broken.

Because of the large number of amino acids (Table 2-1) which nature has to choose from in building enzyme molecules, the properties of an enzyme molecule will depend in large measure on the *sequence* of amino acid residues in the final enzyme molecule. How nature controls the synthesis of a given enzyme so that all its molecules are identical is a very interesting, current field of research; however, that is another story.

TABLE 2-1 The common amino acids

Name	Structure
Glycine (GLY)	$H_3N^+-CH_2-CO_2^-$
Alanine (ALA)	$CH_3-CH(N^+H_3)-CO_2^-$
Valine (VAL)	$(CH_3)_2CH-CH(N^+H_3)-CO_2^-$
Leucine (LEU)	$(CH_3)_2CHCH_2-CH(N^+H_3)-CO_2^-$
Isoleucine (ILEU)	$CH_3CH_2CH(CH_3)-CH(N^+H_3)-CO_2^-$
Serine (SER)	$HOCH_2-CH(N^+H_3)-CO_2^-$
Threonine (THR)	$CH_3CH(OH)-CH(N^+H_3)-CO_2^-$
Cysteine (CYS)	$HSCH_2-CH(N^+H_3)-CO_2^-$
Methionine (MET)	$CH_3SCH_2CH_2-CH(N^+H_3)-CO_2^-$
Aspartic acid (ASP)	$HOC(=O)CH_2-CH(N^+H_3)-CO_2^-$
Asparagine (ASN)	$H_2NC(=O)CH_2-CH(N^+H_3)-CO_2^-$

TABLE 2-1 The common amino acids (*Cont.*)

Name	Structure
Glutamic acid (GLU)	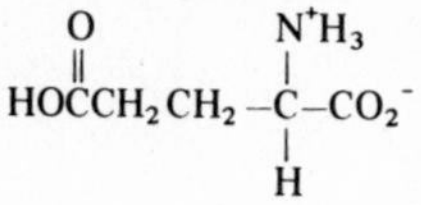 $HO\overset{\overset{O}{\|}}{C}CH_2CH_2-\underset{H}{\overset{N^+H_3}{C}}-CO_2^-$
Glutamine (GLN)	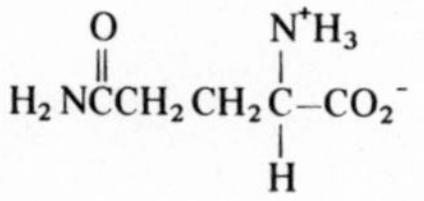 $H_2N\overset{\overset{O}{\|}}{C}CH_2CH_2\underset{H}{\overset{N^+H_3}{C}}-CO_2^-$
Lysine (LYS)	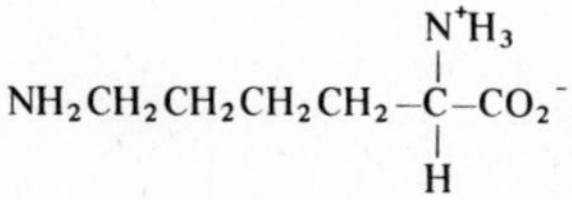 $NH_2CH_2CH_2CH_2CH_2-\underset{H}{\overset{N^+H_3}{C}}-CO_2^-$
Arginine (ARG)	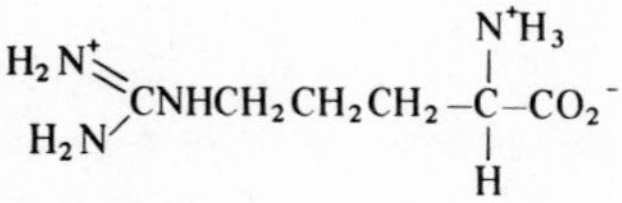 $(H_2N^+{=})(H_2N-)CNHCH_2CH_2CH_2-\underset{H}{\overset{N^+H_3}{C}}-CO_2^-$
Histidine (HIS)	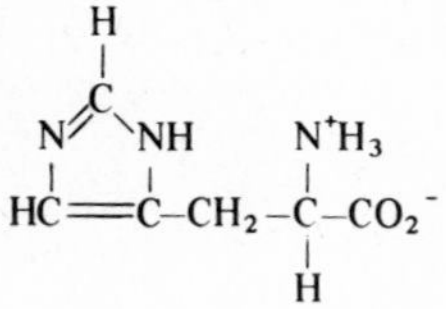 $\overset{H-C(N{=}\ \ NH)}{HC{=}C}-CH_2-\underset{H}{\overset{N^+H_3}{C}}-CO_2^-$
Phenylalanine (PHE)	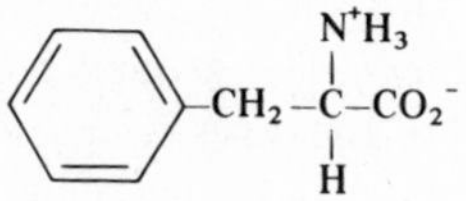 $-CH_2-\underset{H}{\overset{N^+H_3}{C}}-CO_2^-$ (phenyl ring)
Tyrosine (TYR)	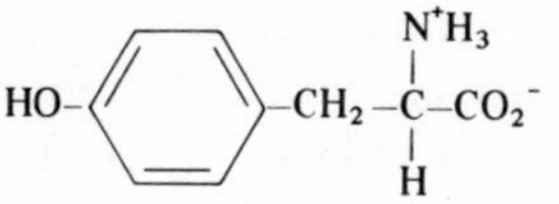 $HO-$(ring)$-CH_2-\underset{H}{\overset{N^+H_3}{C}}-CO_2^-$
Tryptophan (TRY)	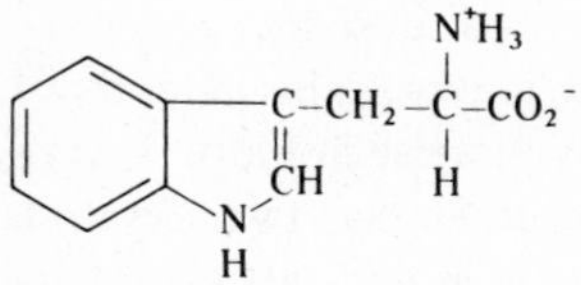 (ring)$\overset{C}{\underset{CH}{\|}}-CH_2-\underset{H}{\overset{N^+H_3}{C}}-CO_2^-$, $\underset{H}{N}$
Proline (PRO)	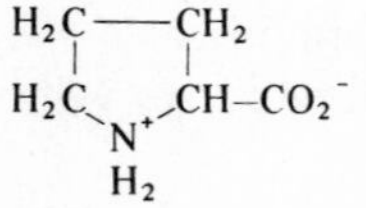 H_2C-CH_2, $H_2C\ \ CH-CO_2^-$, $\underset{H_2}{N^+}$

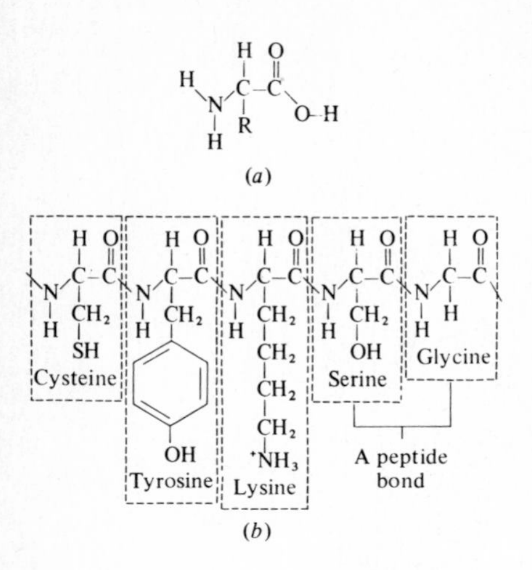

FIGURE 2-1 (*a*) General formula of an amino acid. (*b*) A segment of the amino acid sequence in trypsin. In the native enzyme molecule, the sulfur atom of cysteine is covalently bonded to the sulfur atom of another cysteine residue, thus forming a disulfide bond that helps to maintain the three-dimensional structure of the molecule.

(See, for example, the articles by Clark and Marcker, and by Yanofsky referred to at the end of this chapter.) The sequence of amino acids shown in Fig. 2-1*b* is a small portion of the digestive enzyme trypsin.

The long chain of amino acids which is the enzyme molecule does not flop around freely. Rather it takes up a definite three-dimensional structure. One feature which contributes to the stability of this three-dimensional structure in most enzymes is the presence of disulfide bonds (covalent) between cysteine residues which act as cross-links between different regions of a polypeptide chain or between two polypeptide chains. If the two cysteine residues involved in the disulfide bond are in the same polypeptide chain, the chain forms a sort of "loop." Chymotrypsin has three such loops. It also consists of three polypeptide chains held together by two disulfide bonds. Other relatively weak interactions, such as ionic attraction, hydrogen bonding, and hydrophobic bonding, are important in maintaining the three-dimensional structure of the enzyme. These forces will be discussed in Sec. 2-3.

A purified enzyme consists of molecules which are identical both in amino acid sequence and in three-dimensional structure. A large number of pure enzymes have been prepared as crystalline solids, and in this form they may be studied by the techniques of x-ray crystallography to determine the spatial location of each atom, other than hydrogen, in the enzyme molecule. Such a study is tedious and time-consuming, but very rewarding. To date, the three-dimensional structures of several proteins, including a number of enzymes, have been determined. One such enzyme is lysozyme. For many years chemists have inferred from their studies that enzymes have an "active site," that is, a relatively small region on the enzyme surface where catalysis actually occurs. The x-ray analysis of lysozyme confirms, in vivid detail, the existence of the active site, as can be seen in Fig. 2-2. There is a groove along one side of the molecule into which the substrate molecule fits and is held tightly. Near the center of this groove are two functional groups which are believed to catalytically assist the hydrolysis of the substrate. (Classes of organic compounds are characterized by the presence of certain *functional groups*, e.g., the $-\text{OH}$ group of alcohols and the $-\overset{\text{O}}{\overset{\|}{\text{C}}}-\text{OH}$ group of acids.) The large bulk of the lysozyme molecule, which is not directly involved in the catalytic function of the enzyme, is nonetheless important because its more-or-less rigid, three-dimensional structure maintains the groove and holds the catalytically important functional groups in the correct positions.

Some large enzymes contain more than one active site per molecule. These are often aggregates of identical subunits having molecular weights between 25,000 and 50,000, with one active site per subunit. Most of these enzymes contain four subunits per molecule; however, some examples with other than four subunits are known. Under certain conditions many of these enzymes can be dissociated into their component subunits. Sometimes, but not always, the dissociated subunits remain catalytically active. In other words, aggregation is not always a prerequisite for catalytic activity for these enzymes. Why then do some enzymes exist as aggregates of subunits? One answer we shall explore in Chap. 6 is that an aggregate of subunits is more subject to control than a single subunit.

A word should be said about the naming of enzymes. Although some well-known common names such as chymotrypsin and pepsin are still used, it is now accepted that enzymes should be named systematically and have the ending "-ase." Enzymes have been classified into six general groups: oxidoreductases, transferases, hydrolases, lyases, isomerases, and ligases, which catalyze oxidation-reduction, group transfer, hydrolysis, addition, isomerization, and condensation reactions, respectively.

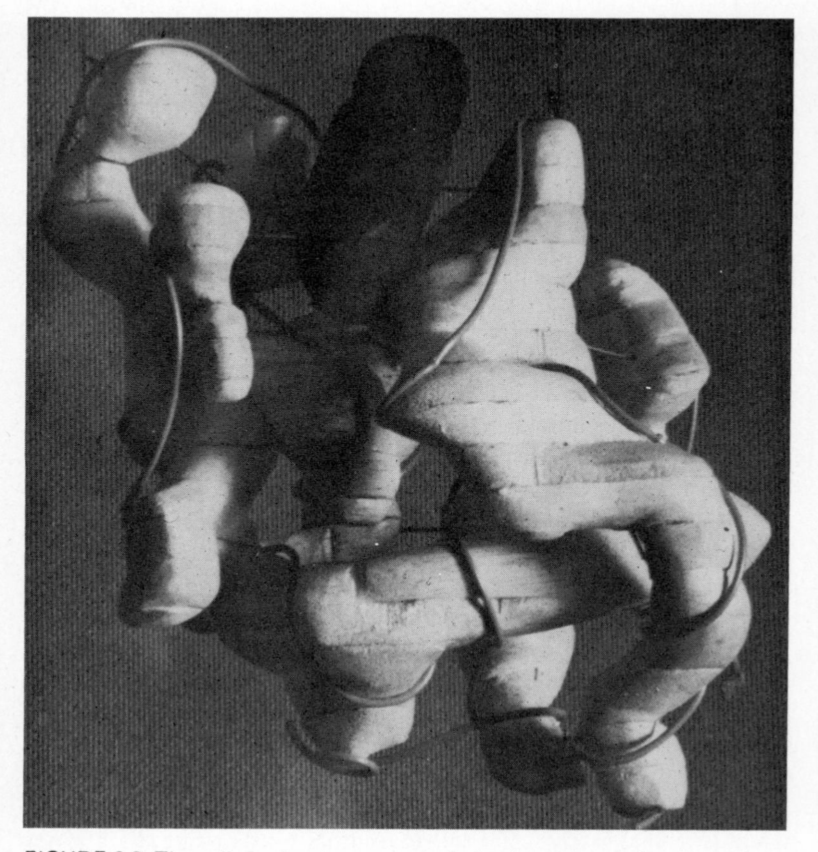

FIGURE 2-2 Three-dimensional structure of the lysozyme molecule as determined by x-ray analysis at 6-Å revolution. The white sausagelike structure encloses the volume that is entirely occupied by the polypeptide chain and its side chains, partly by water. The vertical open groove in front, just left of the middle, is where the substrate lies when it is bound to the enzyme. In this figure, the top half of the groove is filled by a black object that represents an inhibitor molecule. (Figure 3-9 shows the protein structure along the groove in more detail.) [*From C. C. F. Blake et al., "Crystallographic Studies of the Activity of Hen Egg-white Lysozyme," Proc. Roy. Soc. B167, 380 (1967).*]

2-2 COENZYMES

Some enzymes can perform their catalytic function only if certain small (relative to the enzyme) organic molecules are also present. The names of some of these molecules, or *coenzymes*, are listed in Table 2-2 along with the classes of enzymes for which they are coenzymes.

Some of these coenzymes, such as the flavin nucleotides, are so tightly bound to the host enzyme protein that the two together are referred to as the enzyme. Other coenzymes are so loosely held to the enzyme molecule that they readily dissociate and they may properly be thought of as cosubstrates, along with the substrate undergoing reaction. Certain coenzymes alone can catalyze chemical reactions when the normal host enzyme is absent, but with greatly reduced efficiency; however, the coenzyme, by itself, is not a true catalyst since it is changed chemically during the reaction.

The word *substrate* is used to refer to the chemical reactant whose reaction is catalyzed by an enzyme. For example, the enzyme alcohol dehydrogenase, which is found in the liver among other places, catalyzes the conversion of ethyl alcohol to acetaldehyde; here, ethyl alcohol is the substrate.

2-3 THE ENZYME–SUBSTRATE COMPLEX

In all enzyme-catalyzed reactions, the first step is the (noncovalent) binding of the substrate to the enzyme to form what is called the enzyme-substrate complex. This complex can dissociate to give back free substrate and enzyme. Thus, this is an equilibrium step as shown in Eq. (2-1).

$$\underset{\text{Enzyme}}{E} + \underset{\text{Substrate}}{S} \rightleftharpoons \underset{\text{Enzyme-substrate complex}}{ES} \tag{2-1}$$

The rate at which this binding step occurs will depend on the nature of both the enzyme and the substrate; however, in all cases where the binding rate has been measured, it has been found to be very fast, so that the equilibrium of Eq. (2-1) is achieved in a very small fraction of a second.

TABLE 2-2 Some important coenzymes

Coenzyme	Enzymes
Pyridoxal (vitamin B_6)	Transaminases
Thiamine (vitamin B_1)	Decarboxylases, transketolases
Biotin	Carboxylases
Nicotinamide adenine dinucleotide	Dehydrogenases
Flavin nucleotides	Oxidases

We are only beginning to learn what kinds of chemical forces are involved in binding the substrate to the enzyme. Over the years chemists have identified several kinds of intermolecular attractive forces from the study of small molecules, and these same forces are no doubt utilized in the binding of a substrate to an enzyme. One current area of research seeks to determine to what extent each of these forces is important in specific cases. Several types of these forces, and some examples, follow.

1. Ionic attraction. One important factor in the binding of acetylcholine to the enzyme acetylcholinesterase is believed to be the attraction of the positive charge on the nitrogen atom in the substrate for a negatively charged functional group in the enzyme [Eq. (2-2)]. The extremely rapid hydrolysis of acetylcholine in the body is of great importance for the proper functioning of the nervous system.

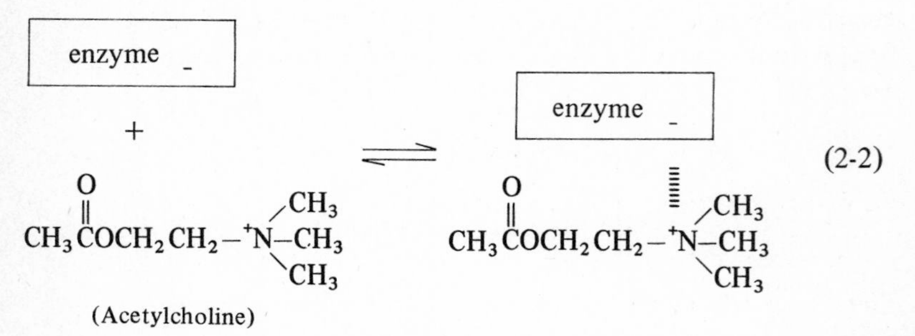

2. Hydrogen bonding. In the digestive tract the hydrolysis of proteins is catalyzed by enzymes such as chymotrypsin. Hydrogen bonding may assist in binding the peptide functional groups of the protein to the enzyme, in a manner suggested by Eq. (2-3).

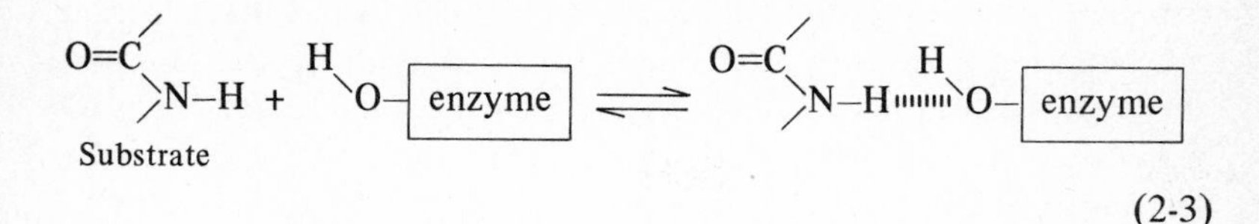

(2-3)

3. Apolar bonding. It is common knowledge that nonpolar substances like gasoline do not dissolve in polar substances like water, but that two nonpolar substances, such as gasoline and oil, are mutually miscible. In the same way, the nonpolar (hydrocarbon) region of a substrate will prefer the nonpolar region of the active site of the enzyme, if there is such a region on the active site, to the polar environment of the water in which both enzyme and substrate are dissolved. Thus the substrate becomes bound by what is called *an apolar interaction* to the enzyme. In the enzyme lysozyme (Fig. 2-2) it is known that the groove into which the substrate fits is largely nonpolar, and this suggests that apolar bonding is of considerable importance in binding substrates to this enzyme. Since life exists in an aqueous environment and many biologically important molecules are nonpolar, apolar bonding is of widespread importance.

The equilibrium between enzyme, substrate, and the enzyme-substrate complex may be characterized mathematically by an equilibrium constant. Biochemists prefer to define the equilibrium constant for the dissociation of the enzyme-substrate complex as shown in Eq. (2-4).

$$\mathrm{ES} \rightleftharpoons \mathrm{E} + \mathrm{S} \qquad K_s = \frac{[\mathrm{E}]\,[\mathrm{S}]}{[\mathrm{ES}]} \tag{2-4}$$

Values of the dissociation constant K_s for many enzyme-substrate complexes are of the order of 10^{-6} to 10^{-3} M. Since enzymes are such efficient catalysts, biochemists usually study reactions in which the concentration of enzyme is much lower than the concentration of substrate. Suppose that $K_s = 1 \times 10^{-5}$ M for a typical enzyme-substrate complex and that a reaction mixture containing 1×10^{-6} M enzyme and 1×10^{-3} M substrate is prepared. We can readily calculate that at equilibrium the concentrations of enzyme, substrate, and enzyme-substrate complex are 0.01×10^{-6}, 0.99901×10^{-3}, and 0.99×10^{-6} M, respectively. In other words, almost all the enzyme has been converted to enzyme-substrate complex; the enzyme is said to be *saturated* with substrate. This is always the case when the substrate concentration is much greater than both K_s [Eq. (2-4)] and the total enzyme concentration. In the preceding example we must keep in mind that the calculated concentrations are correct only at the instant the reaction mixture is prepared because the reaction begins to take place immediately, and this affects the concentrations of the three compounds.

2-4 ENZYME KINETICS

The simplest overall reaction sequence for an enzyme-catalyzed reaction is given in Eq. (2-5), where P stands for the products of the reaction. This mechanism was suggested by the Frenchman Henri in 1903 (although it is usually attributed to Michaelis and Menten whose publication in 1913 is more accessible in the literature) to explain his observations on the rate of hydrolysis of sucrose as catalyzed by the enzyme sucrase (invertase). The enzyme which is regenerated in the final step of Eq. (2-5) immediately reacts with more substrate to form enzyme-substrate complex.

$$\mathrm{E} + \mathrm{S} \underset{k_{-1}}{\overset{k_1}{\rightleftharpoons}} \mathrm{ES} \xrightarrow{k_{\mathrm{cat}}} \mathrm{E} + \mathrm{P} \tag{2-5}$$

A study of the rate, or kinetics, of an enzyme-catalyzed reaction is useful in determining the various steps that occur in the total reaction, i.e., the reaction mechanism. We are now prepared to derive the equation for the rate of the reaction in Eq. (2-5). The constants k_1, k_{-1}, and k_{cat} are rate constants for the reaction in the direction indicated by the associated arrow. The quantities $[\mathrm{E}]_0$ and $[\mathrm{S}]_0$ will denote initial concentrations of enzyme and substrate, and [E], [S], etc., will denote concentrations of enzyme, substrate, etc., at any later time. According to the laws of chemical kinetics and the calculus, we may write the following equations for the rate of change of concentration of substrate ($d[\mathrm{S}]/dt$), of enzyme-substrate complex ($d[\mathrm{ES}]/dt$), and of product ($d[\mathrm{P}]/dt$).

$$-\frac{d[\mathrm{S}]}{dt} = k_1[\mathrm{E}][\mathrm{S}] - k_{-1}[\mathrm{ES}] \tag{2-6}$$

$$\frac{d[\mathrm{ES}]}{dt} = k_1[\mathrm{E}][\mathrm{S}] - k_{-1}[\mathrm{ES}] - k_{\mathrm{cat}}[\mathrm{ES}] \tag{2-7}$$

$$\frac{d[\mathrm{P}]}{dt} = k_{\mathrm{cat}}[\mathrm{ES}] \tag{2-8}$$

Consider again the numerical example in the preceding section. Even though substrate disappears as the reaction progresses, for a considerable part of the reaction the concentration of substrate will still be much greater than the value of K_s and thus the enzyme will remain saturated. In other words, the concentration of the enzyme-substrate complex changes much more slowly than does the concentration of substrate. Thus relative to $d[\mathrm{S}]/dt$, $d[\mathrm{ES}]/dt$ is approximately equal to zero. This is the *steady-state approximation*. Without presenting any proof, we point out that this approximation is valid in general when $[\mathrm{S}] \gg [\mathrm{E}]$, regardless of the value of K_s. With Eq. (2-7) set equal to zero, we can solve for [ES] [Eq. (2-9)].

$$[\mathrm{ES}] = \frac{k_1[\mathrm{E}][\mathrm{S}]}{k_{-1} + k_{\mathrm{cat}}} = \frac{[\mathrm{E}][\mathrm{S}]}{K_m} \tag{2-9}$$

In Eq. (2-9) the constant K_m has been substituted for the collection of rate constants $(k_{-1} + k_{\mathrm{cat}})/k_1$. Of the total concentration of enzyme in solution, part is free enzyme and part is complexed with substrate, i.e.,

$$[\mathrm{E}]_0 = [\mathrm{E}] + [\mathrm{ES}] \quad \text{or} \quad [\mathrm{E}] = [\mathrm{E}]_0 - [\mathrm{ES}] \tag{2-10}$$

[E] from Eq. (2-10) may be substituted into Eq. (2-9) to give

$$[\mathrm{ES}] = \frac{[\mathrm{E}]_0[\mathrm{S}]}{K_m} - \frac{[\mathrm{ES}][\mathrm{S}]}{K_m} \tag{2-11}$$

This may be rearranged to give

$$[\mathrm{ES}] = \frac{[\mathrm{E}]_0[\mathrm{S}]}{K_m + [\mathrm{S}]} \tag{2-12}$$

A consequence of setting $d[\mathrm{ES}]/dt = 0$ is that $-d[\mathrm{S}]/dt = d[\mathrm{P}]/dt$, as may be proved by equating the right-hand sides of Eqs. (2-6) and (2-8) and comparing with Eq. (2-7). Thus the rate of the reaction, v, may be defined either as the rate of disappearance of substrate or the rate of appearance of product. The latter is more convenient, and by substituting Eq. (2-12) in (2-8) we obtain

$$v = \frac{d[\mathrm{P}]}{dt} = \frac{k_{\mathrm{cat}}[\mathrm{E}]_0[\mathrm{S}]}{K_m + [\mathrm{S}]} \tag{2-13}$$

$$K_m = \frac{k_{-1} + k_{\mathrm{cat}}}{k_1} \tag{2-14}$$

The constant K_m, defined earlier and again in Eq. (2-14), is commonly referred to as the Michaelis constant. In the majority of enzyme-catalyzed reactions for which the constants k_1, k_{-1}, and k_{cat} have been determined, $k_{-1} \gg k_{cat}$; in these cases, K_m simplifies to k_{-1}/k_1 and becomes equal to K_s, the dissociation constant for the enzyme-substrate complex.

In deriving Eq. (2-13), we have used the same approach as Briggs and Haldane used in 1925, namely, the steady-state assumption that $d[\mathrm{ES}]/dt$ is negligible compared with $d[\mathrm{S}]/dt$. Henri, on the other hand, started by calculating [ES] from Eq. (2-2). In doing so, he assumed that enzyme and substrate are always in equilibrium, or that $k_{-1} \gg k_{cat}$. His final rate equation was identical to Eq. (2-13) except that K_s replaced K_m. The steady-state assumption is more general than, and therefore preferable to, the equilibrium assumption.

It was pointed out in Sec. 2-3 that when $[\mathrm{S}] \gg K_s$, the enzyme becomes saturated with substrate. Similarly, when $[\mathrm{S}] \gg K_m$, the denominator of Eq. (2-13) becomes approximately equal to [S], and Eq. (2-13) reduces to Eq. (2-15), where V is called the maximum velocity; V is independent of the substrate concentration.

$$V = k_{\mathrm{cat}}[\mathrm{E}]_0 \tag{2-15}$$

Such a reaction is called zero-order since the rate is proportional to the zeroth power of the substrate concentration. In practical terms this means that adding more substrate to the solution will not increase the rate of the reaction, which, according to Eq. (2-15), depends only on the rate constant k_{cat} and the total enzyme concentration.

The other extreme form of Eq. (2-13) occurs when $[\mathrm{S}] \ll K_m$. In this case the velocity is given by

$$v = \frac{k_{\mathrm{cat}}[\mathrm{E}]_0[\mathrm{S}]}{K_m} \tag{2-16}$$

Since the rate of a reaction under these conditions is directly proportional to the substrate concentration raised to the first power, the reaction is first-order in substrate.

A final special case occurs when $[S] = K_m$; then Eq. (2-13) reduces to

$$\nu = \frac{k_{cat}[E]_0}{2} = \tfrac{1}{2}V \tag{2-17}$$

That is, the rate of the reaction is one-half the maximum velocity. This leads to the operational definition of K_m; K_m is the substrate concentration for which the rate is one-half the maximum rate.

The two constants k_{cat} and K_m are important parameters for the reaction of a particular substrate catalyzed by a particular enzyme since they indicate how susceptible the substrate is to catalysis by the enzyme. If, for a particular enzyme, a substrate has a small K_m and a large k_{cat}, its reaction is very rapid in the presence of the enzyme. Such a substrate is said to be a *specific* substrate because it both binds strongly and reacts fast. For example, the alcohols, particularly ethyl alcohol, are specific substrates of liver alcohol dehydrogenase.

When a biochemist has isolated and purified a new enzyme, his usual desire is to determine its specificity, i.e., he sets out to determine the k_{cat} and K_m values for substrates of that enzyme. He does this by measuring the initial rates of the reaction between enzyme and varying concentrations of substrate. If one measures the rate of disappearance of substrate (or the appearance of product) for only the first 5 percent or so of the reaction, the rate is approximately constant since [S] changes only slightly. In Fig. 2-3 are shown the data obtained in the determination of initial rates for the reaction between the enzyme invertase (sucrase) and the substrate sucrose (ordinary sugar) for seven different concentrations of sucrose as determined by Michaelis and Menten in 1913. Since Eq. (2-17) tells us that K_m is the substrate concentration at which the initial rate ν_i is equal to half the maximum rate, we could read off $K_m \approx 0.02\ M$ directly from Fig. 2-3. However, this method introduces considerable error of judgment. Chemists prefer to plot data in such a way that the points lie on a straight line. It is possible to rearrange Eq. (2-13) into the linear form of Eq. (2-18) by taking the reciprocal of both sides, as Lineweaver and Burk pointed out in 1934. Then a plot of $1/\nu_i$ versus $1/[S]_0$, as in Fig. 2-4, should give a straight line with an intercept on the ordinate equal to $1/(k_{cat}[E]_0)$ and a slope of $K_m/(k_{cat}[E]_0)$.

$$\frac{1}{\nu_i} = \frac{1}{k_{cat}[E]_0} + \frac{K_m}{k_{cat}[E]_0}\left(\frac{1}{[S]_0}\right) \tag{2-18}$$

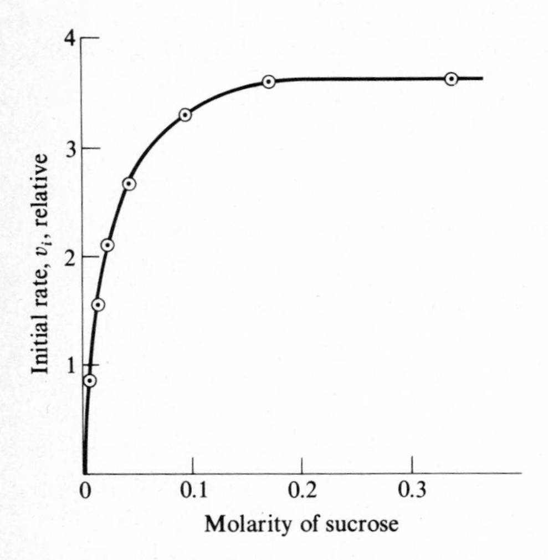

FIGURE 2-3 Initial rates of sucrose hydrolysis at different initial substrate concentrations. Each point represents a different experiment. The enzyme (sucrase) concentration is the same in all experiments. The data used here were obtained by Michaelis and Menten in 1913.

The quotient of the slope divided by the intercept is K_m (= 0.017 *M*); k_{cat} may be calculated from the intercept if $[E]_0$ is known. For the simple hydrolytic enzymes such as chymotrypsin and acetylcholinesterase, and for enzymes containing small cofactors (coenzymes), methods have been developed for determining the enzyme concentration by titration. For other enzymes, $[E]_0$ is very often not known, but relative values of k_{cat} may be determined by choosing one substrate as a reference and comparing other substrates to it.

Up to this point we have been considering the relatively simple mechanism for enzyme catalysis shown in Eq. (2-5). However, most enzymatic reactions are more complicated. Many of the metabolic enzymes must bind a coenzyme or a second substrate in addition to the primary substrate. Then, too, there is generally more than one step in the reaction of the enzyme-substrate complex. In spite of these complications, the *form* of the rate equation in Eq. (2-13) is generally applicable, and this is why we have considered the oversimplified mechanism of Eq. (2-5) in so much detail. For a more complicated mechanism the constants k_{cat} and K_m do not refer to a single step; they become composite constants incorporating the rate constants for several steps or equilibria.

The mechanism of action of several hydrolytic enzymes is shown in Eq. (2-19). An example is chymotrypsin which catalyzes the hydrolysis of proteins and of simple carboxylic acid derivatives such as esters and amides. A typical simple substrate is the ethyl ester of the amino acid tyrosine.

$$\mathrm{E} + \mathrm{S} \underset{k_{-1}}{\overset{k_1}{\rightleftharpoons}} \mathrm{ES} \xrightarrow{k_2} \underset{+\ \mathrm{P}_1}{\mathrm{ES}'} \xrightarrow{k_3} \mathrm{E} + \mathrm{P}_2 \tag{2-19}$$

As shown in Eq. (2-19), this substrate first binds with the enzyme to form the enzyme-substrate complex which then undergoes internal reaction to release ethyl alcohol P_1 and to form the acyl-enzyme ES′. The acyl-enzyme is a new ester in which the carboxylic acid part of the original ester has reacted with the alcohol functional group of a particular serine residue in the enzyme molecule, as shown in Eq. (2-20).

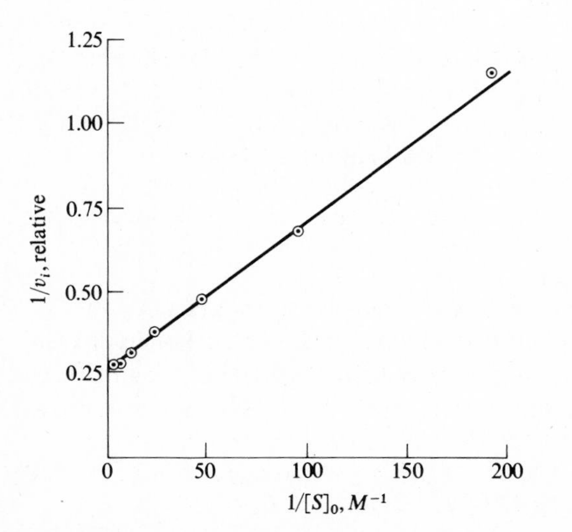

FIGURE 2-4 A Lineweaver-Burk plot of the data from Fig. 2-3 for the hydrolysis of sucrose catalyzed by the enzyme sucrase. Applying Eq. (2-18), a value for K_m of 0.017 M is calculated.

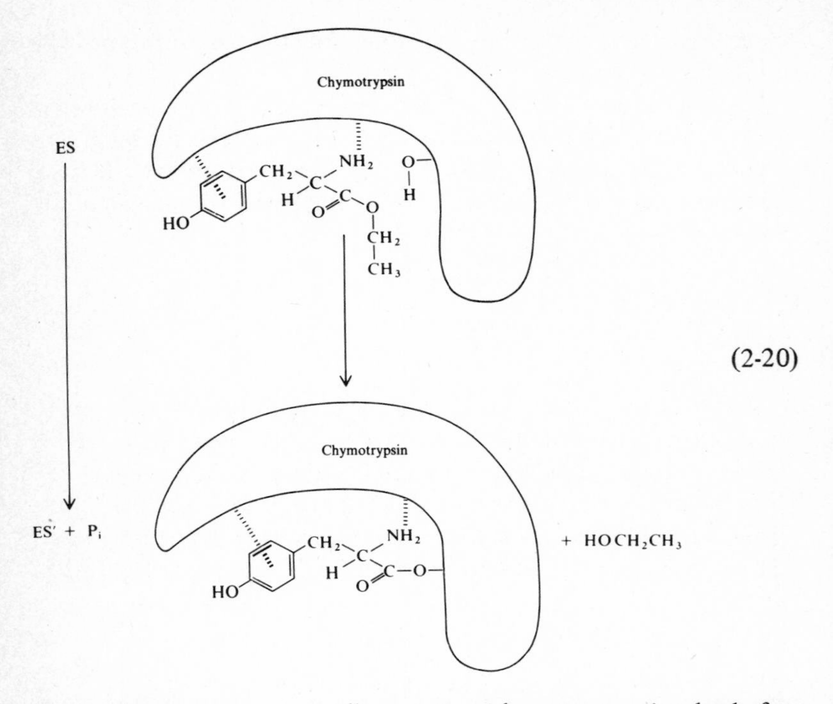

(2-20)

The acyl-enzyme very rapidly reacts with water to give back free enzyme and tyrosine P_2. The rate equation in Eq. (2-13) still applies to the mechanism of Eq. (2-19) and the methods described above for determining the constants k_{cat} and K_m may still be used, but the constants have the new definitions shown in Eq. (2-21).

$$k_{cat} = \frac{k_2 k_3}{k_2 + k_3} \qquad K_m = \frac{k_3}{k_2 + k_3} \frac{k_{-1} + k_2}{k_1} \tag{2-21}$$

Thus k_{cat} is composed of the two first-order rate constants k_2 and k_3, and K_m is a function of all the rate constants. These relationships can often be simplified, and the reader is urged to determine what k_{cat} and K_m reduce to in the very common case where $k_{-1} \gg k_2$ and $k_2 \gg k_3$. (Recall that $k_{-1}/k_1 = K_s$.)

2-5 ENZYME SPECIFICITY

As noted above, a specific substrate is one which has a small K_m value and large k_{cat}. Thus an enzyme exhibits two types of specificity toward substrates, binding specificity as reflected in the K_s values, and kinetic specificity as reflected in the k_{cat} values. Enzymes exhibit varying degrees of specificity. At one extreme is the hydrolytic enzyme urease which is almost absolutely specific because it has only one really good substrate, urea. Many enzymes are specific in the sense that they catalyze a particular type of reaction. For example, chymotrypsin catalyzes the hydrolysis of proteins, and also of amides and esters of amino acids. However, not all the amino acid esters are equally susceptible to catalysis by chymotrypsin. It is much more specific for the derivatives of the aromatic amino acids, tyrosine, phenylalanine, and tryptophan, than of the other amino acids. Aromatic compounds are those which contain the six-carbon phenyl ring (see tyrosine structure in Fig. 2-1) or a similar ring structure. Apparently the active site of chymotrypsin is of such a nature that (1) it has a strong attraction for aromatic groups in compounds (binding specificity), and (2) it holds the bound aromatic substrate in just the right orientation for other groups at the active site to catalyze the breaking of old covalent bonds and the making of new covalent bonds (kinetic specificity).

Chymotrypsin is also typical of many enzymes which exhibit stereochemical specificity. Each of the amino acids (except glycine) can exist as two different stereoisomers which are called *optical* isomers because they are capable of rotating the plane of polarized light either to the left or to the right. The two isomers differ in the arrangement (configuration) of the different parts of the molecule about the asymmetric carbon atom, as shown for alanine in Fig. 2-5; consequently they rotate polarized light in opposite directions. The two isomers are mirror images, i.e., the isomer on the right has the same configuration as the form on the left appears to have when viewed in a mirror. (The circle in Fig. 2-5 represents a carbon atom which is said to be asymmetric because the four groups bonded to it are all different. A hydrogen atom extends below the page, and the other three groups shown extend above the page in accordance with the tetrahedral geometry about the carbon atom.) The amino acids found in nature are almost always of the L configuration, i.e., their R– groups are directed in the same way as the CH_3– group in L-alanine.

If a substrate of chymotrypsin, say tyrosine ethyl ester, is synthesized from a mixture of D- and L-tyrosine, only the hydrolysis of L-tyrosine ethyl ester will be catalyzed by chymotrypsin. Thus chymotrypsin shows stereochemical specificity. It should be added that although the isomer with the D configuration does not react, it forms

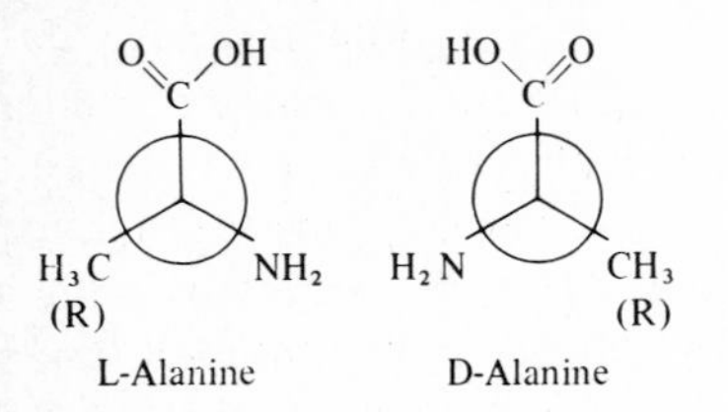

FIGURE 2-5 Optical isomers of the amino acid alanine. The circle represents the asymmetric carbon atom to which four different groups are bonded. A hydrogen atom points out from it below the page. The other three groups shown point slightly up out of the page in the usual tetrahedral geometry. Other amino acids than alanine have their characteristic side chains (R) in the position of the methyl group.

just as stable an enzyme-substrate complex (ES) with chymotrypsin as does the L isomer; thus it must be that the D isomer is bound in the wrong orientation for the catalytic groups at the active site. (See Chap. 6.)

2-6 ENZYME INHIBITION

An enzyme inhibitor is a substance which reduces, or completely eliminates, the ability of an enzyme to act as a catalyst. There are two general classes of inhibitors: (1) *reversible* inhibitors, which do not form covalent bonds with the enzyme and which allow the enzyme to regain complete activity when removed from solution; and (2) *irreversible* inhibitors, which form covalent bonds with the enzyme and generally reduce or destroy enzyme activity permanently. We shall consider one type of inhibition, which occurs at the active site, from each of these two classes.

Competitive inhibition occurs when a substance competes with the normal substrate at the active site of an enzyme. It forms an enzyme-inhibitor complex which is in (reversible) equilibrium with free enzyme and free inhibitor in the same way as a substrate forms an enzyme-substrate complex. Such an inhibitor has an equilibrium binding constant K_i, defined in the same way as K_s.

$$K_i = \frac{[E][I]}{[EI]} \qquad (2\text{-}22)$$

It is reasonable that competitive inhibitors are very often structurally similar to specific substrates since they bind to the enzyme at the same place as the substrates do. For example, the ethyl ester of D-tyrosine, mentioned above in the discussion of enzyme specificity, is a competitive inhibitor of chymotrypsin. Another classic example is the competitive inhibition of the enzyme succinate dehydrogenase by malonate ion, which is structurally similar to the succinate ion, the specific substrate for this enzyme. (See Fig. 2-6.)

In a reaction mixture containing enzyme, specific substrate, and a competitive inhibitor, some of the enzyme will be in the form of enzyme-inhibitor complex and therefore unavailable as a catalyst for the substrate. Thus it is obvious in a qualitative way why the reaction will be slower in the presence of the inhibitor than in its absence. Mathematically, the equation for the rate of the reaction when a competitive inhibitor is present is given in Eq. (2-23), where [I] is the concentration of inhibitor.

$$v = \frac{k_{cat}[E]_0[S]}{[S] + K_m(1 + [I]/K_i)} \tag{2-23}$$

Some concrete cases will breathe life into Eq. (2-23). When $[S] = K_m$ and $[I] = K_i$, the reader should prove to his own satisfaction that $v = V/3$. In contrast, when $[S] = K_m$ and no inhibitor is present, $v = V/2$ according to Eq. (2-23). Thus under these conditions the inhibited reaction is only two-thirds as fast as the uninhibited reaction. However, if the substrate concentration is made considerably greater than K_m while $[I] = K_i$, the rate of reaction attains the same maximum rate as in

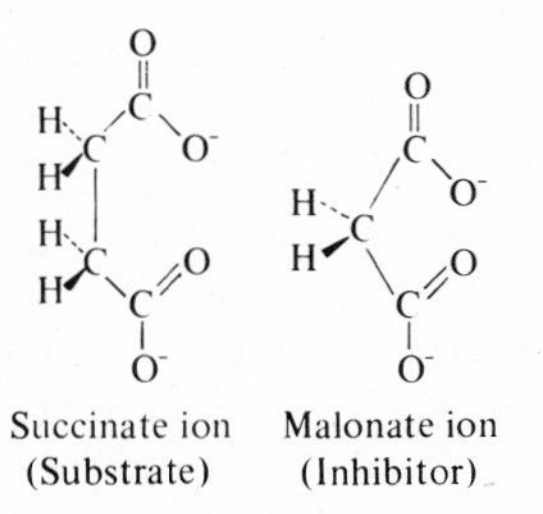

FIGURE 2-6 An example of a competitive inhibitor. The malonate ion (inhibitor) is structurally similar to the succinate ion that is the substrate for the enzyme succinate dehydrogenase.

the absence of inhibitor. Thus under these conditions the inhibitor is not inhibiting the reaction. Why is this? According to Eq. (2-22), when $[I] = K_i$, $[E] = [EI]$, that is, the amount of enzyme tied up in the form of enzyme-inhibitor complex is the same as the amount of *free* enzyme. But when $[S] \gg K_m$ the amount of free enzyme is very small compared to the amount of enzyme-substrate complex. Thus the amount of enzyme-inhibitor complex is also very small; hence almost all the enzyme is in the form of enzyme-substrate complex and the rate is the maximum rate.

Since the product of an enzyme-catalyzed reaction is very often structurally similar to the substrate in the reaction, the product can sometimes act as a competitive inhibitor. As such a product accumulates, it will gradually slow down the reaction. Product inhibition is one of the ways in which the chemical reactions in a living organism are controlled.

The second type of inhibition we shall consider is irreversible inhibition in which the inhibitor reacts at the active site to form a stable covalent bond which is not readily broken. The most dramatic inhibitors of this type include several of the poisonous "gases." The nerve gases, such as diisopropylphosphorofluoridate, are potent irreversible inhibitors of the enzyme acetylcholinesterase. These organophosphorus compounds form a stable covalent bond between the phosphorus atom and the hydroxyl oxygen atom of the important serine residue in the enzyme [Eq. (2-24)].

```
  CH3 O
   |  ||
H-C-O-P-F      +    HO-[enzyme]   ---->
   |   |
  CH3  O
       |
  CH3-C-CH3
       |
       H
   Diisopropyl       Acetylcholinesterase
phosphorofluoridate

  CH3 O
   |  ||
H-C-O-P-O-[enzyme]   +  HF      (2-24)
   |   |
  CH3  O
       |
  CH3-C-CH3
       |
       H
  Enzyme-inhibitor
     compound
```

The nerve gases are deadly because when the enzyme acetylcholinesterase is inactivated, acetylcholine accumulates in the synapses and interferes with the transmission of nerve impulses. Thus the brain loses control over vital bodily functions, such as breathing, and paralysis and death result.

Although an irreversibly inactivated enzyme cannot be reactivated simply by removing the excess free inhibitor from the presence of the enzyme, certain chemicals can reactivate such an enzyme by chemically reacting with the inhibitor fragment which is covalently bonded to the enzyme. Such an antidote for the nerve gases is 2-pyridine aldoxime methiodide, or 2-PAM, which is highly reactive towards the phosphorus atom that is bonded to the serine oxygen atom of the inhibited enzyme.

Besides acetylcholinesterase, several of the proteolytic enzymes such as trypsin and chymotrypsin are also inhibited by the nerve gases by reaction at the important serine residue in the active site. This particular reaction was a clue which led to the elucidation of the mechanism of action of these enzymes [Eq. (2-19)]. Like the nerve-gas inhibitors, substrates also react with the enzyme to form an acyl-enzyme intermediate in which the carboxylic acid, or acyl, part of the substrate is bonded to a serine oxygen atom on the enzyme. In the case of a good substrate this intermediate, which is really an ester in which the enzyme supplies the alcohol part, reacts readily with water to regenerate free enzyme and release the carboxylic acid part of the original substrate. The facility of the reaction of this intermediate is what distinguishes a good substrate from irreversible inhibitors of the nerve-gas type. With several substrates of low specificity, the acyl-enzyme intermediate reacts sufficiently slowly that it can be separated from the reaction mixture and crystallized as a pure compound. For example, the rather poor substrate of chymotrypsin, *p*-nitrophenyl trimethylacetate, forms the intermediate trimethylacetyl-chymotrypsin which has been isolated in crystalline form. In solution this acyl-enzyme is completely inactive, but when the trimethylacetyl group comes off by reaction with water, complete enzyme activity is regained. A study of inhibitors and substrates of low specificity is useful in determining the mechanism of enzyme-catalyzed reactions, as illustrated by the discovery of the acyl-enzyme intermediate in the case of chymotrypsin-catalyzed reactions.

By studying the various aspects of enzyme behavior that have been discussed in this chapter—kinetics, specificity, inhibition, structure, and the role of coenzymes–chemists have learned a great deal about the mechanism of enzyme action.

SUGGESTED READINGS

Michaelis, L., and M. L. Menten: The Kinetics of Invertase (Sucrase) Action, *Biochemische Zeitschrift,* vol. 49, p. 333, 1913. This original paper will be of interest to those who have a reading knowledge of German. Much of the paper presents experimental results. The interesting part from the historical point of view is the introductory part (pp. 333-336) in which Henri's prior work is acknowledged, and the rate equation is derived (pp. 343–344).

Watson, J. D.: "Molecular Biology of the Gene," 2d ed., W. A. Benjamin, Inc., New York, 1970. This book is an excellent introduction to biology in molecular terms. Chapter 4 discusses weak interactions, such as those forces which determine enzyme structure, and the binding of substrates to enzymes.

Stein, W. H., and S. Moore: The Chemical Structure of Proteins, *Scientific American,* p. 81, February, 1961. The methods still in use today for determining amino acid sequences in proteins are described by the two men who developed the techniques for separating and quantitatively measuring amino acids. Although several new sequence-determining techniques have been introduced since 1961, the basic approach is still the same.

Perutz, M. F.: The Hemoglobin Molecule, *Scientific American,* p. 64, November, 1964. The principles and techniques of x-ray analysis are described, along with their application to the determination of the three-dimensional structure of the hemoglobin molecule in the crystalline state. More recently the same author has published the hemoglobin structure at 2.8 Å resolution in *Nature,* vol. 219, pp. 29 and 131, 1968.

Clark, B. F. C., and K. A. Marcker: How Proteins Start, *Scientific American,* p. 36, January, 1968. Reviews the general sequence of events in protein biosynthesis with major emphasis on how the process starts.

Yanofsky, C.: Gene Structure and Protein Structure, *Scientific American,* p. 80, May, 1967. A discussion of how genes control the amino acid sequence in proteins and some of the steps involved in protein synthesis.

Segal, H. L.: The Development of Enzyme Kinetics, in "The Enzymes," 2d ed., vol. 1, Academic Press, Inc., New York, 1959. A detailed historical account of the development of the concept of an enzyme-substrate complex and of enzyme kinetics. Although it is a scholarly work, much of the first 30 pages or so should be comprehensible to the first-year student.

THREE
CATALYSIS INVOLVING ACIDS AND BASES

Catalysis by acids and bases is extremely common in chemical reactions. Even the complex biochemical catalysts, enzymes, utilize catalysis by acids and bases. In this chapter we shall explore acid-base catalysis. We shall start with some simple examples of different types of catalysis by acids and bases and proceed to consider how enzymes utilize these same principles.

3-1 ACIDS, BASES, AND PROTON TRANSFERS

What is an acid and what is a base? Of the many answers chemists have given to this question, the most convenient definitions for our purposes were given by Professors Brønsted in Denmark and Lowry in England in 1923. They defined an acid as a proton donor, i.e., a substance which can *donate* a proton (or hydrogen ion, H^+) to another substance. A base, on the other hand, is a proton acceptor, i.e., a substance which can *accept* a proton from another substance.

Implicit in these definitions is the assumption that the proton, or hydrogen ion, does not exist as a separate chemical species, but is always bonded to some other molecule or ion. Thus in water, an acid such as HCl does not really dissociate into a H^+ and a Cl^-; rather, it transfers a proton to a water molecule. In reality the H_3O^+ ion (hydronium ion), is also an oversimplified representation of the true state of affairs since it is known to be associated very closely with at least three other water molecules. In Eq. (3-1) we see that, according to the Brønsted definitions, HCl is an acid since it is a proton donor; H_2O is a base since it is a proton acceptor.

$$HCl + H_2O \longrightarrow H_3O^+ + Cl^- \tag{3-1}$$

Hydrochloric acid is such a strong acid in water that essentially all HCl molecules lose their protons to water molecules as indicated by the single arrow in Eq. (3-1). However, in an aqueous solution of a weak acid only a small fraction of the weak acid molecules donate their protons to water at any given instant. Equation (3-2) shows the reaction of the weak acid, acetic acid, with water.

$$CH_3\overset{\overset{\displaystyle O}{\|}}{C}OH + H_2O \underset{k_{-1}}{\overset{k_1}{\rightleftharpoons}} CH_3\overset{\overset{\displaystyle O}{\|}}{C}O^- + H_3O^+ \qquad (3\text{-}2)$$

This is an equilibrium reaction, and an equilibrium constant, called the *acid dissociation constant* K_a, may be defined in the usual way.

$$K_a = \frac{[CH_3\overset{\overset{\displaystyle O}{\|}}{C}O^-][H_3O^+]}{[CH_3\overset{\overset{\displaystyle O}{\|}}{C}OH]} \qquad (3\text{-}3)$$

The amount of water which is protonated to give H_3O^+ is exceedingly small compared with the total amount of water molecules in ordinary dilute solutions. Therefore, the concentration of water is not significantly changed by the partial dissociation of acetic acid, and the $[H_2O]$ term has been incorporated into the equilibrium constant K_a. The reaction in Eq. (3-2) is in dynamic equilibrium, i.e., both forward and reverse reactions occur at equal rates. In the forward reaction an acetic acid molecule donates a proton to a water molecule; in the reverse reaction the hydronium ion donates a proton to the acetate ion. Thus, in the reverse reaction, hydronium ion is an acid and the acetate ion is a base according to Brønsted's definitions. The acetate ion is said to be the conjugate base of acetic acid; conversely, acetic acid is the conjugate acid of the acetate ion. The acetic acid–acetate ion pair is called a conjugate acid-base pair. Similarly, the hydronium ion-water pair is a conjugate acid-base pair.

In dilute aqueous acetic acid solution, the concentration of acetic acid is considerably greater than the concentration of hydronium ion; hence we can infer that acetic acid is a weaker acid than the hydronium ion. The concentrations of the solute species in Eq. (3-3) can be determined experimentally. The value $K_a = 1.8 \times 10^{-5}\ M$ has been calculated in this way. Table 3-1 contains a list of weak acids and their K_a values. The quantity pK_a is the negative logarithm of K_a.

In recent years chemists have developed techniques for measuring chemical reactions which are extremely fast. The rate of the reaction in the forward direction in Eq. (3-2) is given mathematically by

$$\text{Rate} = k_1 [\text{CH}_3\overset{\overset{\text{O}}{\|}}{\text{C}}\text{OH}][\text{H}_2\text{O}] \tag{3-4}$$

where "Rate" refers to the rate of decrease in concentration of acetic acid in moles per liter per second ($M \cdot s^{-1}$). Since the two concentration terms on the right-hand side of Eq. (3-4) have the units of molarity (M), the units of the rate constant will be $M^{-1} \cdot s^{-1}$. The experimental value for k_1 is $1.5 \times 10^4\ M^{-1} \cdot s^{-1}$. In a dilute solution of acetic acid the concentration of water is almost the same as in pure water, namely 55 M, and it changes negligibly during the dissociation of acetic acid; i.e., the reaction is pseudo-first-order, proceeding at a rate dependent only on the concentration of acetic acid. Therefore, we may rewrite Eq. (3-4) to give

$$\begin{aligned} \text{Forward rate} &= 1.5 \times 10^4\ M^{-1} \cdot s^{-1} \times 55\ M \times [\text{CH}_3\overset{\overset{\text{O}}{\|}}{\text{C}}\text{OH}] \\ &= 8.2 \times 10^5\ s^{-1} \times [\text{CH}_3\overset{\overset{\text{O}}{\|}}{\text{C}}\text{OH}] \end{aligned} \tag{3-5}$$

The number $8.2 \times 10^5\ s^{-1}$ is the pseudo-first-order rate constant for the dissociation of acetic acid. A first-order rate constant has a simple reationship to $t_{1/2}$, the half-life for the reaction, as shown in Eq. (3-6).

$$t_{1/2} = \frac{0.693}{k_{\text{first order}}} \tag{3-6}$$

The value of $t_{1/2}$ for acetic acid in water is thus 8.5×10^{-7} s, which means that if the reverse reaction in Eq. (3-2) did not occur, half of the acetic acid molecules would have donated their protons to water molecules within approximately one millionth of a second after the acid was added to the water. In fact, since the reverse reaction does occur, equilibrium is achieved in a considerably shorter time.

Referring to Eq. (1-14), we may write

$$K_a = \frac{k_1}{k_{-1}} \times 55\ M \tag{3-7}$$

Since we know k_1 and K_a, we can calculate that $k_{-1} = 4.5 \times 10^{10}$ $M^{-1} \cdot s^{-1}$. Thus the reverse reaction is even faster than the forward reaction which "explains" why acetic acid is a weak acid. In fact the reverse reaction is so fast that it is not limited by the free energy of activation for the reaction; $\Delta G^{\ddagger}$ is so small that even at room temperature virtually all collisions are sufficiently energetic to result in a reaction. The limiting factor is the rate at which an acetate ion and a hydronium ion can diffuse together in the solution and collide. Such reactions are said to be diffusion-controlled.

As the preceding examples illustrate, proton transfer reactions are generally very fast. This means that a catalytic mechanism involving proton transfer can potentially be a very efficient type of catalysis. Generally, however, some factor other than the rate of proton transfer will limit the reaction rate in a catalyzed reaction involving a proton transfer.

3-2 CATALYSIS BY HYDRONIUM IONS

As an example of a reaction catalyzed by hydronium ions we shall consider the hydrolysis of an acetal. An acetal is a compound in which two RO– groups are bonded to the same carbon atom as shown by the following general formula:

```
    OR'
    |
R—C—OR"
    |
    R'''
```

Compounds of the acetal type, such as starch and other polysaccharides, are of biological importance. Sucrose is a simple example. It is a disaccharide composed of two monosaccharides, glucose and fructose. The carbon atom indicated by the dotted arrow in Eq. (3-8) has two RO– groups bonded to it.

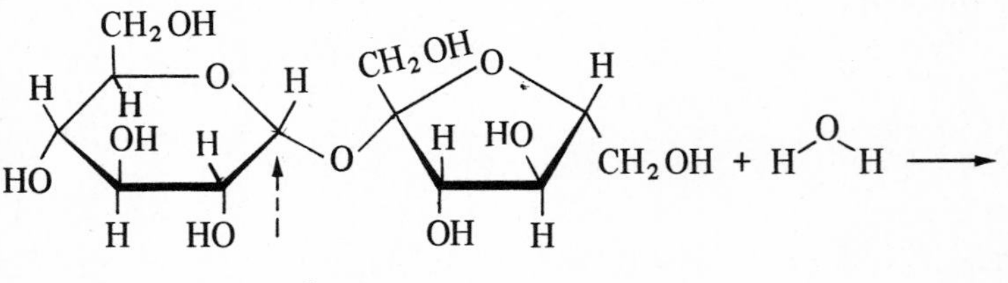

Sucrose

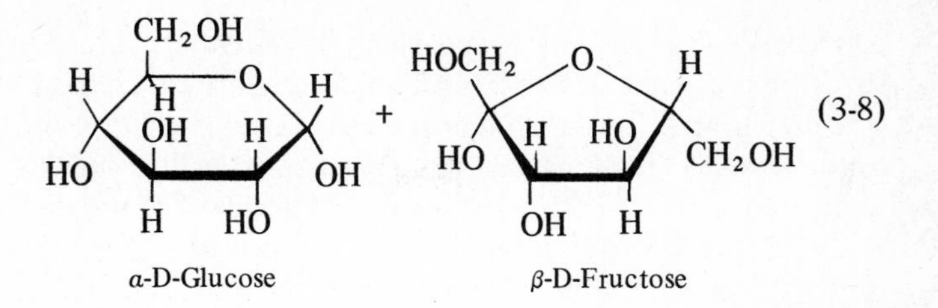

α-D-Glucose β-D-Fructose

When sucrose hydrolyzes, as shown in Eq. (3-8), the bond between the acetal carbon and the (nonring) oxygen atom breaks, and a molecule of water adds to the two fragments. The net result is that two bonds break, those indicated by the arrows in Eq. (3-9), and two new bonds form, as indicated by the dotted lines in Eq. (3-9). (Only the part of the sucrose molecule which reacts is shown.)

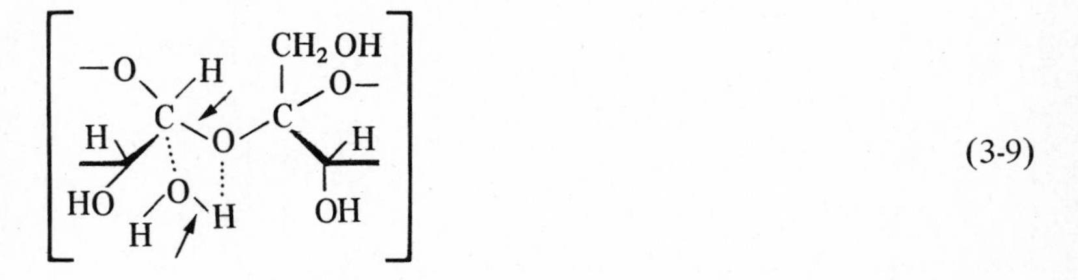

The simultaneous breaking and forming of these four bonds requires considerable energy and does not occur readily. Instead, the reaction probably occurs in four steps, one step for each bond that is made or broken. The slowest of these steps determines the rate of the reaction. The hydronium ion catalyzes the reaction by bypassing the slow step via a new pathway which requires less energy. The slowest step in this reaction is the first one, the breaking of the bond between carbon and oxygen in the sucrose molecule.

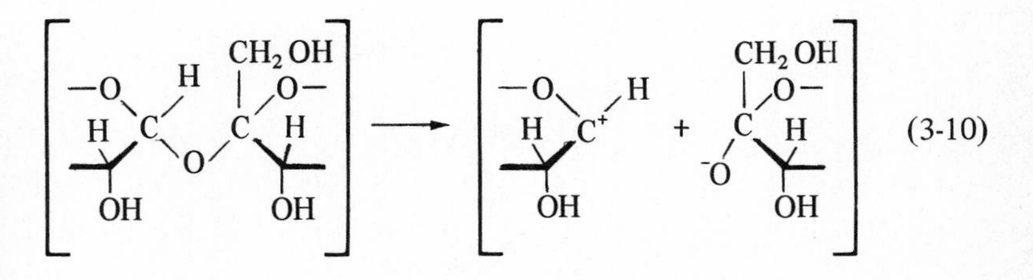

The hydronium ion catalyzes the hydrolysis of sucrose by providing a slightly different pathway for the breaking of the C–O bond, a pathway of considerably lower energy. The hydronium ion (produced in solution by adding a strong acid such as hydrogen chloride) transfers a proton to the oxygen atom of sucrose as shown in Eq. (3-11), and the carbon-oxygen bond in this new protonated molecule breaks much more readily than in the unprotonated molecule. The product fragment in Eq. (3-12) in which the carbon atom is positively charged reacts very rapidly with a molecule of water and the product of that reaction transfers a proton to another water molecule.

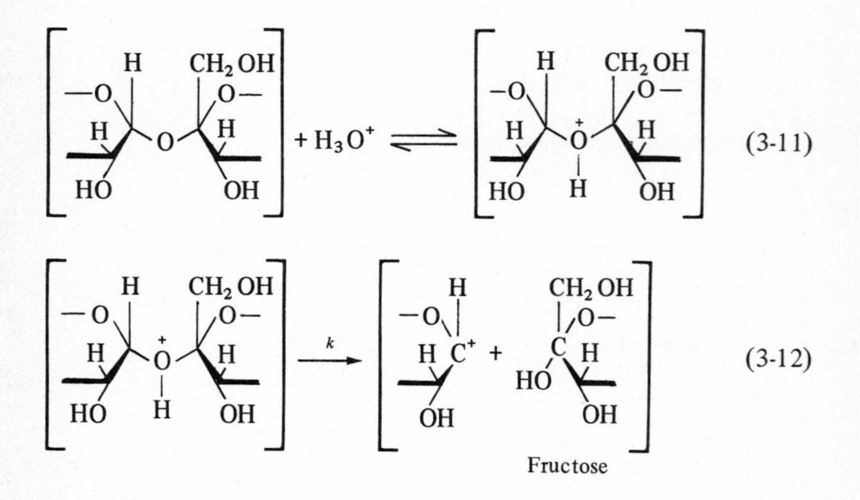

Fructose

These two final steps are shown in Eq. (3-13).

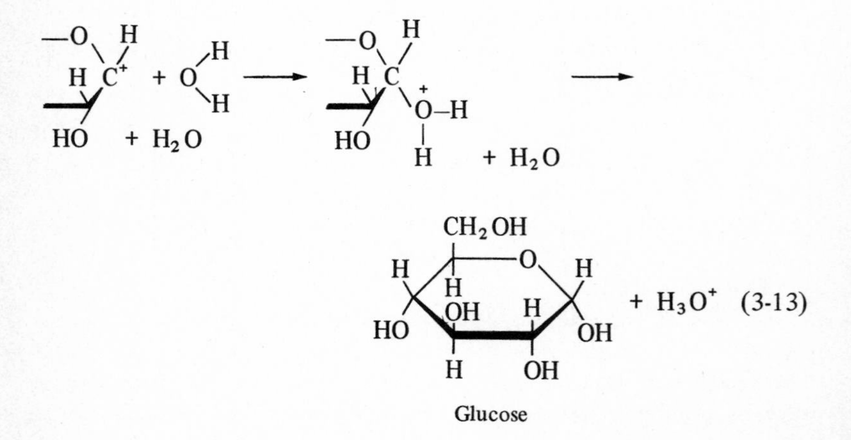

Glucose

If one adds together Eqs. (3-11), (3-12), and (3-13), the net reaction is the same as Eq. (3-8). Thus the hydronium ion catalyst is not consumed in this hydrolytic reaction.

In the catalyzed hydrolysis of sucrose, as in the uncatalyzed reaction, the breaking of the carbon-oxygen bond [Eq. (3-12)] is still the slow, or rate-determining, step. Consequently, the overall rate of the reaction is given by Eq. (3-14), where SH^+ is used as a symbol for the protonated molecule of sucrose in Eq. (3-12). From the equilibrium constant K_a, for the reverse reaction in Eq. (3-11), it is possible to derive a relationship between sucrose concentration [S] and protonated sucrose concentration. This relationship may be substituted into Eq. (3-14) to give Eq. (3-16). Thus the rate of the hydrolysis of sucrose is proportional to the concentration of sucrose [S] and also to the concentration of hydronium ion.

$$\text{Rate of disappearance of sucrose} = k[SH^+] \tag{3-14}$$

$$K_a = \frac{[S][H_3O^+]}{[SH^+]} \qquad \therefore [SH^+] = \frac{[S][H_3O^+]}{K_a} \tag{3-15}$$

$$\text{Rate of disappearance of sucrose} = \frac{k}{K_a}[S][H_3O^+] \tag{3-16}$$

In other words, the more catalyst that is present, the faster the reaction goes. This can be understood in terms of Le Chatelier's principle, since the concentration of protonated sucrose will be increased as the concentration of hydronium ion is increased, and increasing the latter results in an increased rate of reaction according to Eq. (3-14). Even so, the concentration of protonated sucrose is very small unless very concentrated solutions of acid are used.

The reason that we observe hydronium ion catalysis in the hydrolysis of sucrose is that the carbon-oxygen bond in protonated sucrose breaks much more readily [Eq. (3-12)] than in unprotonated sucrose [Eq. (3-10)]. Why is this? In protonated sucrose the (negatively charged) electron pair, which constitutes the bond between the carbon and oxygen atoms, tends to spend more time with the positively charged oxygen atom in protonated sucrose than with the neutral oxygen atom in unprotonated sucrose. Consequently, the process of breaking the carbon-oxygen bond, which involves a complete transfer of that electron pair to the oxygen atom [Eq. (3-12)], has already partly occurred in protonated sucrose. Thus, the final breaking of the bond will take only a small additional amount of energy compared with the energy required to break the carbon-oxygen bond in the unprotonated sucrose molecule.

3-3 FROM EXPERIMENT TO MECHANISTIC CONCLUSIONS

We have just discussed a chemical reaction in considerable mechanistic detail. It is worthwhile at this point to pause a moment to discuss how kinetic results can lead to conclusions about the mechanism of a reaction.

Most chemical reactions proceed through several steps as mentioned in the preceding example. The slowest of these steps determines the rate of the reaction and is called the *rate-determining step*. If only one substrate molecule is involved in the rate-determining step, the reaction is first-order, i.e.,

$$\text{Rate} = k_0[\text{S}] \tag{3-17}$$

where k_0 is the first-order rate constant. If a catalyst exists for the reaction, it must participate in, and accelerate, the rate-determining step, or else it must provide an alternative, and faster, pathway.

In any reaction at a given catalyst concentration, part of the substrate will react via the uncatalyzed pathway, the rest via the catalyzed pathway. The catalyzed reaction has a rate given by

$$\text{Rate} = k_{\text{cat}}[\text{catalyst}]^n[\text{S}] \tag{3-18}$$

where k_{cat} is the catalytic rate constant, and n represents the number of catalyst molecules that participate in the rate-determining step. For the simplest case involving only one catalyst molecule, $n = 1$. The total rate of reaction is given by adding Eqs. (3-17) and (3-18) to give

$$\text{Rate} = (k_0 + k_{\text{cat}}[\text{catalyst}]^n)[\text{S}] \tag{3-19}$$

If the catalyst is not consumed during the reaction, its concentration remains constant. Then the reaction will be pseudo-first-order, since the rate in Eq. (3-19) depends only on the substrate concentration, i.e.,

$$\text{Rate} = k_{\text{obs}}[\text{S}] \tag{3-20}$$

where k_{obs} is the measured pseudo-first-order rate constant.

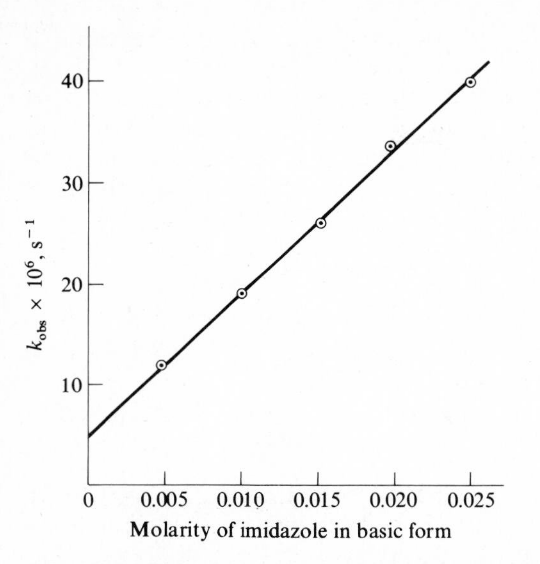

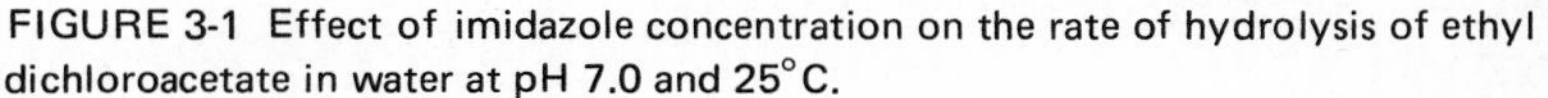

FIGURE 3-1 Effect of imidazole concentration on the rate of hydrolysis of ethyl dichloroacetate in water at pH 7.0 and 25°C.

$$k_{obs} = k_0 + k_{cat}[\text{catalyst}]^n \quad (3\text{-}21)$$

A useful experimental procedure is to study the rate of the reaction at varying catalyst concentrations. When k_{obs} is plotted as a function of catalyst concentration, the relationship is linear, as shown in Fig. 3-1 and expressed in Eq. (3-21) for $n = 1$. Such a result yields three pieces of information: (1) the relative efficiency of the catalyst as reflected in the value of k_{cat}, the slope of the straight line in Fig. 3-1; (2) the rate of the reaction in the absence of catalyst as reflected in k_0, the intercept of the straight line in Fig. 3-1; and (3) the number of catalyst molecules involved in the rate-determining step—one in this case, since k_{obs} is observed to be linearly dependent on the *first* power of the catalyst concentration.

3-4 MECHANISM OF HYDROLYSIS OF CARBOXYLIC ACID DERIVATIVES

In organic chemistry certain atoms or groups of atoms, which are found in many different compounds and which retain their distinctive chemical characteristics, are called functional groups. The —OH group in alcohols and the $-NH_2$ groups in amines are common examples. The carboxyl group ($-C\overset{\displaystyle /\!\!/ O}{\underset{\displaystyle \backslash OH}{}}$), found in the carboxylic acids, is another example. In the remainder of this chapter we shall be discussing several examples of hydrolysis reactions of compounds derived from the carboxylic acids. The general formulas for the most common of these derivatives are given here.

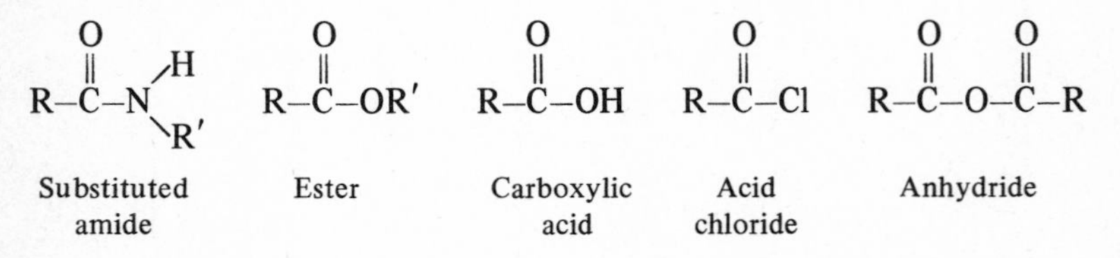

As before, the letter R symbolizes an organic chain of varying length and composition. All these derivatives hydrolyze in water according to the overall reaction given in Eq. (3-22).

$$R-\overset{\overset{\displaystyle O}{\|}}{C}-X + H_2O \longrightarrow R-\overset{\overset{\displaystyle O}{\|}}{C}-OH + HX \qquad (3\text{-}22)$$

The general formula $R-C\overset{\displaystyle /\!\!/ O}{\underset{\displaystyle \backslash X}{}}$ refers to any derivative of a carboxylic acid. The rates of reaction vary considerably; amides hydrolyze very slowly, anhydrides quite rapidly. The detailed mechanism for the hydrolysis of these compounds is given in Eqs. (3-23) and (3-24).

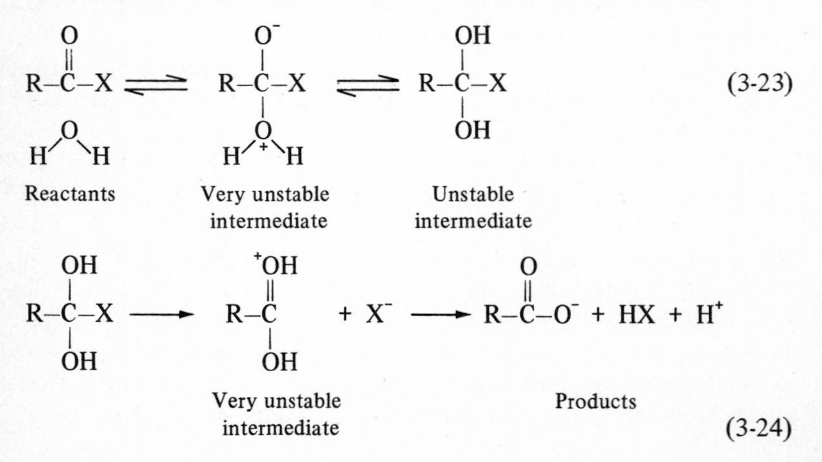

The first step in Eq. (3-23) is called an addition reaction because a molecule of water "adds" to the substrate. This is usually the rate-determining step of the reaction. The addition compound formed in this step may revert back to the initial reactants as indicated by the reverse arrow. The second step in Eq. (3-23) [and also the second step in Eq. (3-24)] is really several steps involving proton transfers. The positively charged oxygen atom probably does not transfer its proton directly to the negatively charged oxygen atom, but rather to an adjacent water (solvent) molecule; similarly, the negatively charged oxygen atom will probably get its proton from a water molecule near it. These proton transfers between oxygen atoms are usually extremely fast and will not be illustrated in detail in writing mechanisms.

The structures shown in Eqs. (3-23) and (3-24), except for the reactants and products, are called intermediates. They are generally quite unstable and either revert to reactants or react rapidly via the next step. Thus their concentrations are generally very small. Of the three intermediates illustrated, the last one in Eq. (3-23) is the most stable. It is commonly called a tetrahedral intermediate because the four groups attached to the carbon atom are almost tetrahedrally disposed about it. In contrast both reactant and product have three groups in a planar arrangement about this carbon atom. The tetrahedral intermediate reacts rapidly as shown in Eq. (3-24), losing the X group and undergoing several rapid proton transfers to give the final products of the reaction. At neutral pH the carboxylic acid will be in the anionic form as shown. Adding Eqs. (3-23) and (3-24) together gives the overall balanced equation for the hydrolysis shown in Eq. (3-22).

3-5 GENERAL ACID CATALYSIS

In Sec. 3-2 we discussed the hydronium ion-catalyzed hydrolysis of sucrose. The reader may have wondered at that point whether acids other than H_3O^+ could catalyze the reaction by transferring a proton to the substrate. For that particular reaction the answer is "no." A strong acid such as hydrogen chloride catalyzes the hydrolysis of sucrose, not because it transfers a proton directly to the sucrose molecule in aqueous solution, but because it reacts with water almost completely to form hydronium ions which do catalyze the reaction. Any other strong acid is an equally good catalyst. Thus the ability of a given acid to catalyze the hydrolysis of sucrose does not depend on the identity of the acid per se, but on its ability to create hydronium ions in the solution.

However, many reactions are susceptible to catalysis by proton transfer from acids other than (and including) the hydronium ion. Such reactions are for this reason said to be general acid-catalyzed. One such reaction which we shall discuss briefly is the hydrolysis of amides. The amides are of interest here because the amino acids in proteins are joined together by amide bonds. The overall equation for the hydrolysis of acetamide in neutral (pH 7) aqueous solution is shown in Eq. (3-25).

$$\underset{\text{Acetamide}}{CH_3\overset{\overset{O}{\|}}{C}-NH_2} + H_2O \longrightarrow \underset{\text{Acetate ion}}{CH_3\overset{\overset{O}{\|}}{C}-O^-} + NH_4^+ \qquad (3\text{-}25)$$

Since the reaction is exceedingly slow at room temperature, it has been studied at temperatures of 100°C or higher. In the presence of weak acids it is found that the reaction is accelerated to an extent proportional to the concentration of added weak acids. Therefore, the weak acid must participate in some way in the rate-determining step of the reaction. Although the detailed mechanism for the general acid-catalyzed hydrolysis of amides has not been definitely established, one possible mechanism is shown in the following equations, where BH symbolizes any weak acid.

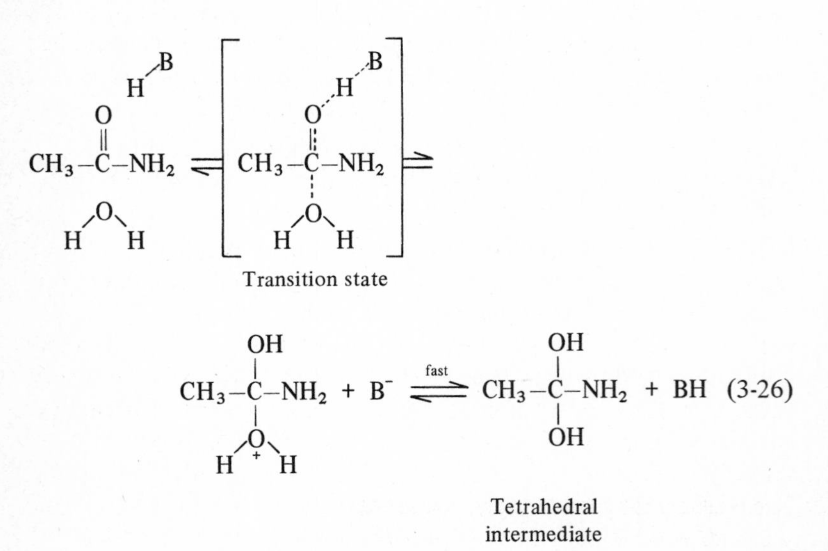

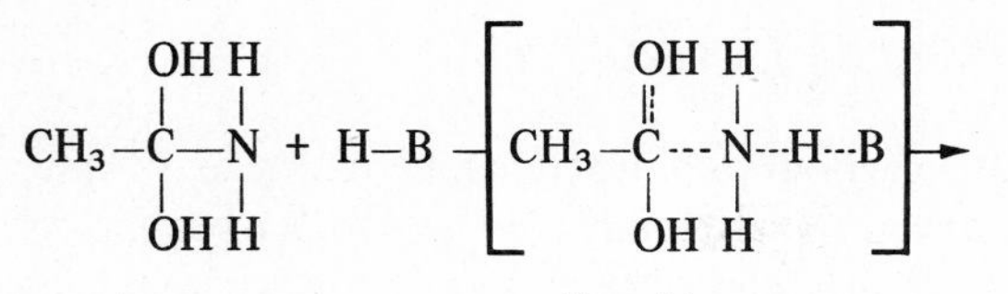

Transition state

$$CH_3-\overset{\overset{+}{O}H}{\overset{\|}{C}} \underset{OH}{} + NH_3 + B^- \xrightarrow{\text{fast}} CH_3\overset{O}{\overset{\|}{C}}-O^- + NH_4^+ + BH \qquad (3\text{-}27)$$

This mechanism is very similar to the mechanism for the uncatalyzed reaction discussed in the previous section. The important difference is that the weak acid assists in the rate-determining step which is the first step of Eq. (3-26), the addition of water to the substrate. The transition state for this step is shown in the square brackets contained within the arrow for the reaction. The dotted lines represent bonds that are either breaking or forming. The proton which is transferred to the substrate during this step performs much the same catalytic function as described in Sec. 3-2; by drawing electrons away from the carbonyl carbon atom, the proton enhances the ability of that carbon atom to accept a share in a pair of electrons donated by the oxygen atom of the attacking water molecule. In the language of Chap. 1, the transfer of a proton in the rate-determining step decreases the free energy of activation for the formation of the bond between a water molecule and the carbonyl carbon atom. Consequently, a larger fraction of the collisions which occur each second between amide and water molecules are sufficiently energetic to result in reaction to give a tetrahedral intermediate. Therefore, the rate of the rate-determining step, and of the overall reaction, has been increased. The acid is thus a catalyst.

The second step in Eq. (3-26) involves very rapid proton transfers as discussed in the previous section. The further reaction of the tetrahedral intermediate is shown in Eq. (3-27). The first step of this equation is probably also assisted by the general acid catalyst. The final reshuffling of protons again occurs very rapidly.

3-6 GENERAL BASE CATALYSIS

General base-catalyzed reactions are reactions in which any base, including hydroxide ion, can act as a catalyst by accepting a proton. As an example of this type of reaction we shall consider the hydrolysis of ethyl dichloroacetate, an ester. The overall reaction is shown in Eq. (3-28).

$$\mathrm{Cl{-}\underset{\displaystyle Cl}{\overset{\displaystyle H}{C}}{-}\overset{\displaystyle O}{\overset{\|}{C}}{-}O{-}CH_2CH_3 + H_2O \longrightarrow Cl_2CH\overset{\displaystyle O}{\overset{\|}{C}}{-}O^- + HOCH_2CH_3 + H^+} \tag{3-28}$$

In the absence of catalysts the reaction proceeds by the general mechanism described earlier in Eqs. (3-23) and (3-24). A variety of weak bases catalyze the reaction. A typical base is imidazole which is available as a white crystalline solid and dissolves readily in water. Its structural formula is shown below.

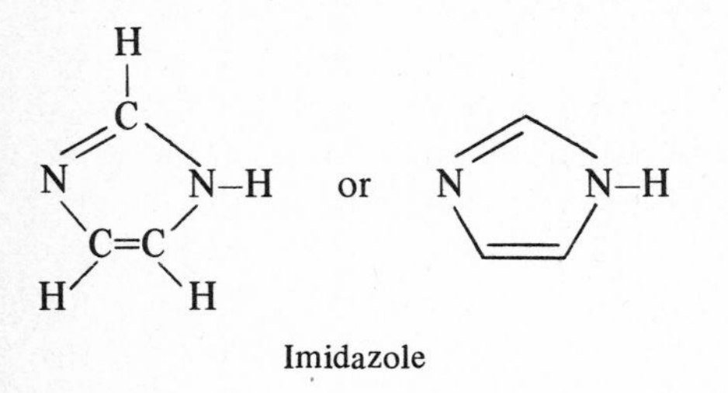

Imidazole

A plot of k_{obs} versus imidazole concentration for this reaction at varying concentrations of imidazole gives a straight line (Fig. 3-1). As explained in Sec. 3-3, we learn from this plot that $k_0 = 5.0 \times 10^{-6}\ \mathrm{s}^{-1}$, $k_{\mathrm{cat}} = 1.4 \times 10^{-3}\ M^{-1} \cdot \mathrm{s}^{-1}$, and one molecule of imidazole participates in the rate-determining step of the reaction. To give the reader some feeling for these numbers, a useful exercise is to calculate the concentration of imidazole at which half of the substrate reacts via the catalyzed pathway and half via the uncatalyzed pathway; i.e., what is the imidazole concentration when the two terms on the right side of Eq. (3-21) are equal? The reader is further urged to satisfy himself that at this imidazole concentration the half-life of the reaction is 69,000 s.

In the uncatalyzed hydrolysis of ethyl dichloroacetate, the rate-determining step is the first step in Eq. (3-23), the addition of water to the ester. Therefore the imidazole molecule catalyzes the reaction by accepting a proton in this step. The probable mechanism for this step is shown in Eq. (3-29). The tetrahedral intermediate produced in this first step, after accepting a proton at the negatively charged oxygen atom, reacts rapidly to give products according to the mechanism of Eq. (3-24).

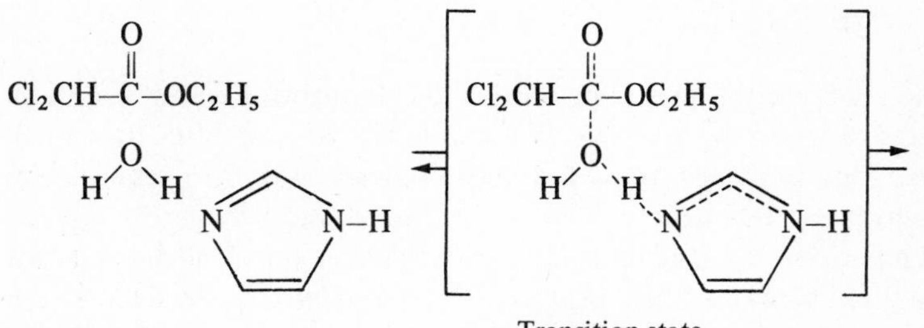

Transition state

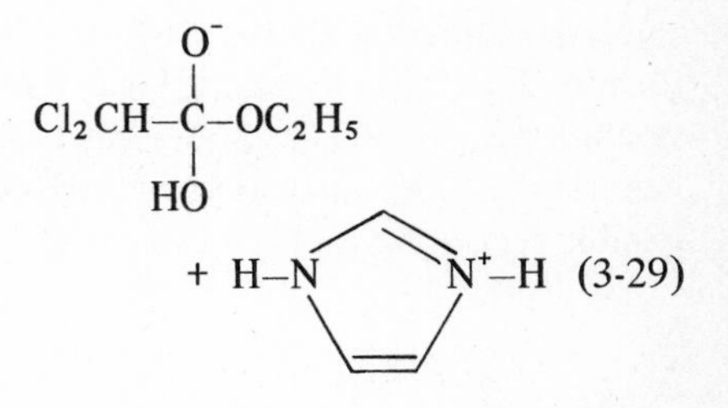

We may now ask the question: why is the reaction in Eq. (3-29) accelerated by the participation of a base? In the transition state the base accepts a proton from the attacking water molecule. As a consequence, the oxygen atom of the water molecule begins to acquire a greater share of the electrons that constituted the bond with the departing proton. Thus the oxygen atom of the water molecule becomes more negatively charged, and its ability to donate an electron pair to form a new bond with the carbonyl carbon atom is enhanced. This greater reactivity is offset, but only partly, by the lesser probability that the ester will collide with a water molecule hydrogen-bonded to the base (catalyzed reaction) than with a free water molecule (uncatalyzed reaction). The net result is that the presence of the base lowers the free energy of activation.

It is of interest to compare general acid and general base catalysis of the hydrolysis of carboxylic acid derivatives. In both types of reactions a bond must be formed between the oxygen atom of a water molecule and the carbonyl carbon atom of the substrate. A general base makes the formation of this bond easier by activating the oxygen atom of water with extra electrons by pulling a proton from it. A general acid makes the formation of this bond easier by making the carbonyl carbon atom deficient in electrons and thus more susceptible to attack.

3-7 THE BRØNSTED CATALYSIS LAW

While studying the preceding section the thoughtful reader may have guessed that strong bases are better general base catalysts than weak bases. This is indeed true, and the discovery was first reported by Brønsted in Denmark in 1924 in his classical research on the decomposition of nitramide. He found that k_B, the catalytic constant for a given base (used here instead of k_{cat}), is related to the dissociation constant of the conjugate acid of that base as shown in Eq. (3-30), where G and β are constants. This is the Brønsted law for general base catalysis. Equation (3-30) may be converted to the more convenient form of Eq. (3-31) by taking logarithms of both sides. In Eq. (3-31) the symbol pK_a is equal to $-\log(K_a)$. According to Eq. (3-31) $\log k_B$ increases as pK_a increases. A series of acids with increasing pK_a's exhibit decreasing acidity (Table 3-1).

$$k_B = G_B\left(\frac{1}{K_a}\right)^{\beta} \qquad (3\text{-}30)$$

$$\log k_B = \log G_B + \beta(pK_a) \qquad (3\text{-}31)$$

However, the conjugate bases of this series of acids show increasing base strength. Therefore, Eq. (3-31) expresses the observed fact that $\log k_B$ increases as base strength increases. This relationship is illustrated by data for the hydrolysis of ethyl dichloroacetate. (See Sec. 3-6.) Catalytic rate constants for several bases have been determined. A plot of $\log k_B$ versus pK_a for some of these data is given in Fig. 3-2. The points in this plot fall on a straight line, and according to Eq. (3-32), the slope of this line is equal to β. Thus $\beta = 0.47$ or approximately one-half. Consequently, for two bases which differ 100

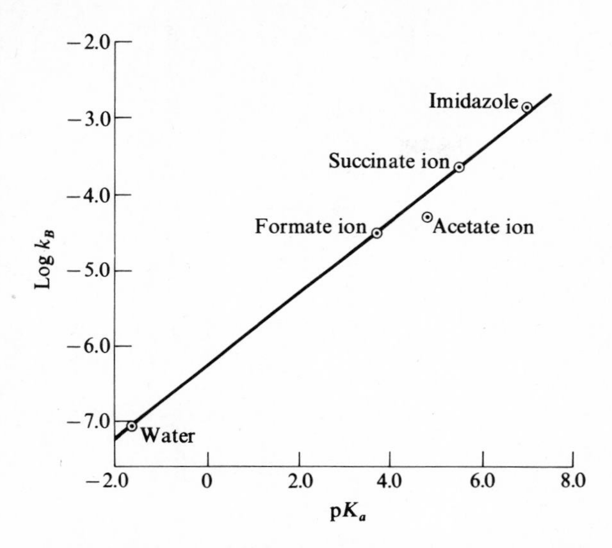

FIGURE 3-2 A Brønsted plot for the general base-catalyzed hydrolysis of ethyl dichloroacetate at 25°C.

times in base strength (two pK_a units), the corresponding catalytic rate constants differ approximately 10 times (one logarithmic unit).

The question arises whether water itself might be acting as a general base catalyst in the hydrolysis of ethyl dichloroacetate; i.e., might a second molecule of water accept a proton from the attacking water molecule in Eq. (3-29) in the so-called "uncatalyzed" reaction? If so, k_0 would be equal to $k_B[H_2O]$ where $[H_2O] \approx 55\ M$. The value of k_B calculated in this way for water lies on the same straight line as the other general bases in Fig. 3-2. Therefore, a second water molecule probably does act as a base catalyst (proton acceptor) in the "uncatalyzed" hydrolysis of ethyl dichloroacetate.

3-8 NUCLEOPHILE CATALYSIS OF THE HYDROLYSIS OF CARBOXYLIC ACID DERIVATIVES

Although we shall deal more extensively with nucleophile catalysis in Chap. 5, we briefly consider here the role of nucleophile catalysts in the hydrolysis of carboxylic acid derivatives. The word nucleophile is a general term referring to any chemical compound or ion which

TABLE 3-1 Ionization constants of some weak acids

Acid	Ionization Reaction	pK_a
Formic acid	$H_2O + H{-}C({=}O){-}OH \rightleftharpoons H{-}C({=}O){-}O^- + H_3O^+$	3.7
Acetic acid	$H_2O + CH_3C({=}O){-}OH \rightleftharpoons CH_3C({=}O){-}O^- + H_3O^+$	4.8
p-Nitrophenol	$H_2O + O_2N{-}C_6H_4{-}OH \rightleftharpoons O_2N{-}C_6H_4{-}O^- + H_3O^+$	7.0
Dihydrogen phosphate ion	$H_2O + HO{-}P({=}O)(OH){-}O^- \rightleftharpoons HO{-}P(O)(O^-){-}O^- + H_3O^+$	7.0
Imidazole	$H_2O + H{-}N^+(C_3H_2)N{-}H \rightleftharpoons N(C_3H_2)N{-}H + H_3O^+$	7.0
Ammonium ion	$H_2O + NH_4^+ \rightleftharpoons NH_3 + H_3O^+$	9.2
Phenol	$H_2O + C_6H_5{-}OH \rightleftharpoons C_6H_5{-}O^- + H_3O^+$	10.0

β-Carboxyl group of aspartic acid†	$H_2O + R{-}C({=}O){-}OH \rightleftharpoons R{-}C({=}O){-}O^- + H_3O^+$	4.1
γ-Carboxyl group of glutamic acid†	$H_2O + R{-}C({=}O){-}OH \rightleftharpoons R{-}C({=}O){-}O^- + H_3O^+$	4.5
Imidazole group of Histidine†	$H_2O + H{-}N\langle\text{ring, R}\rangle N^+{-}H \rightleftharpoons H{-}N\langle\text{ring, R}\rangle N + H_3O^+$	7.0
Thiol group of cysteine†	$H_2O + R{-}SH \rightleftharpoons R{-}S^- + H_3O^+$	8.5
Hydroxyl group of tyrosine†	$H_2O + R{-}C_6H_4{-}OH \rightleftharpoons R{-}C_6H_4{-}O^- + H_3O^+$	9.6
ε-Amine group of lysine†	$H_2O + R{-}N^+H_3 \rightleftharpoons R{-}NH_2 + H_3O^+$	10.2
Guanidino group of arginine†	$H_2O + R{-}NH{-}C(NH_2){=}N^+H_2 \rightleftharpoons R{-}NH{-}C(NH_2){=}NH + H_3O^+$	12.0

† Acidic groups in proteins.

contains an atom that has an unshared pair of electrons. A good nucleophile is one in which the nucleophilic atom readily shares this pair of electrons with an electron-deficient (or potentially electron-deficient) atom of another molecule to form a new covalent bond. Bases fit into the category of nucleophiles because in accepting a proton the base shares a pair of electrons with it to form a new bond. By sharing their electron pairs with atoms other than protons, bases can act as nucleophiles as well as proton acceptors.

In our previous discussion of the addition of water to various compounds (Secs. 3-2 and 3-4 to 3-6), water acted as a nucleophile. Another important nucleophile is the hydroxide ion. In Sec. 3-6 of this chapter we observed that a water molecule assisted by a general base is a better nucleophile than the unassisted water molecule. The hydroxide ion, a water molecule with the proton completely removed, is an even better nucleophile. When a hydroxide ion is substituted for the water molecule in Eqs. (3-23) and (3-24), the reactions lead directly to the same hydrolysis products. Since the hydroxide ion is consumed in this reaction, it is perhaps better to speak of it as a promoter rather than a catalyst.

Many other nucleophiles can react with derivatives of carboxylic acids. This opens up the possibility of catalysis by nucleophiles. The example of bromide ion catalysis of methyl iodide hydrolysis in Chap. 1 is an example of nucleophile catalysis in the hydrolysis of alkyl halides.

In the field of carboxylic acid derivatives, an example of nucleophile catalysis is the hydrolysis of acetic anhydride as catalyzed by formate ion shown in Eqs. (3-32) and (3-33). In the absence of formate ion, acetic anhydride would hydrolyze according to Eqs. (3-23) and (3-24). However, formate ion (added to the solution by dissolving sodium formate) reacts with acetic anhydride [Eq. (3-32)] much faster than water does.

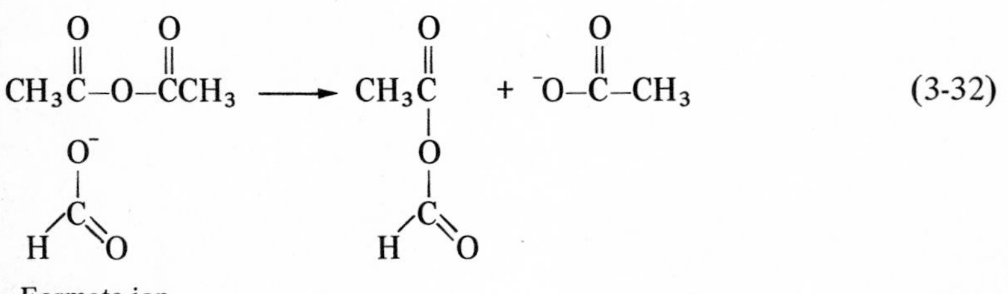

$$CH_3\overset{O}{\overset{\|}{C}}{-}O{-}\overset{O}{\overset{\|}{C}}{-}H + H_2O \longrightarrow CH_3\overset{O}{\overset{\|}{C}}{-}O^- + 2H^+ + H{-}\overset{O}{\overset{\|}{C}}{-}O^- \qquad (3\text{-}33)$$

Furthermore, the reaction in Eq. (3-33) between water and the unsymmetrical anhydride produced in Eq. (3-32) is also faster than the reaction of water with acetic anhydride. Thus formate ion accelerates the hydrolysis of acetic anhydride without being consumed; therefore, it is a catalyst. In addition it is a *nucleophile* catalyst, because in Eq. (3-32) it provides the pair of electrons to form a new bond giving a reactive unsymmetrical anhydride intermediate. The reader should bear in mind that Eq. (3-32) is not a detailed mechanism; a tetrahedral intermediate is formed, no doubt analogous to Eq. (3-23), but with formate ion replacing the water molecule as a nucleophile. Equation (3-33) proceeds also by the two steps shown in Eqs. (3-23) and (3-24). The reader will find it a useful exercise to write out these transition states and intermediates.

In the formate ion-catalyzed hydrolysis of acetic anhydride, the normal rate of hydrolysis of acetic anhydride is increased by an amount proportional to the concentration of formate ion in solution. That is, Eq. (3-21) is obeyed. From this information alone, one cannot tell whether formate ion is acting as a general base catalyst or as a nucleophile catalyst. Other experimental evidence is necessary to distinguish between the two possible types of catalysis. In this particular case, the formate ion is 20 times better as a catalyst than the acetate ion. Acetate ion cannot act as a nucleophile catalyst because the reaction analogous to Eq. (3-32), in which acetate ion is substituted for formate, does not give a reactive intermediate; the reaction gives acetyl exchange, and the products of the reaction are chemically identical with the reactants. Such acetyl exchange does occur, since radioactivity is incorporated into acetic anhydride when it is dissolved in the presence of acetate ion which is radioactively labeled with ^{14}C.

$$CH_3\overset{\overset{O}{\|}}{C}{-}O{-}\overset{\overset{O}{\|}}{C}{-}CH_3 + CH_3{}^{14}\overset{\overset{O}{\|}}{C}O^- \rightleftharpoons CH_3{}^{14}\overset{\overset{O}{\|}}{C}{-}O{-}\overset{\overset{O}{\|}}{C}CH_3 + CH_3\overset{\overset{O}{\|}}{C}O^- \tag{3-34}$$

Since acetate ion cannot act as a nucleophile catalyst, it must be a general base catalyst. According to the Brønsted catalysis law, if formate ion were also a general base catalyst, it should be a poorer catalyst than acetate ion because it is less basic than acetate ion. Since formate ion is observed to be a much better catalyst than acetate ion, it is probably a nucleophile catalyst.

This type of procedure for distinguishing general base from nucleophile catalysis works only in certain special cases. A comparative study of the effectiveness of a catalyst in H_2O (ordinary water) and D_2O (heavy water) serves as a more universal approach. In general base catalysis in H_2O, the general base accepts a proton (H^+) from an H_2O molecule in the transition state; in the same reaction in D_2O, the general base accepts a deuteron (D^+) from a D_2O molecule in the transition state [Eq. (3-35)].

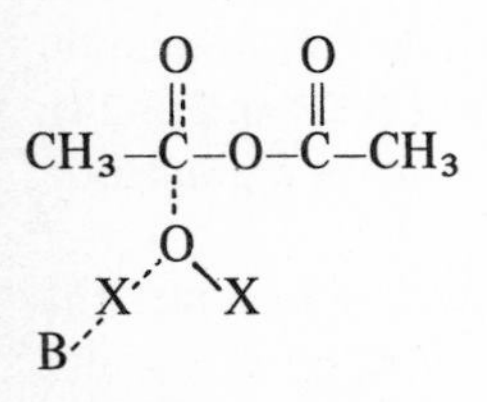

(3-35)

where X = H or D

B = general base

Since the deuterium atom (also D^+) is twice as heavy as the hydrogen atom, the D–O bond in D_2O is stronger than the H–O bond in H_2O, according to quantum mechanics. Consequently, although this bond is only partly broken in the transition state, the energy required for a deuteron transfer is expected to be slightly higher than that for a proton transfer. This expectation is borne out experimentally. For reactions which are general base-catalyzed, the catalytic rate constant for a given base is on the order of two to three times smaller in D_2O than in H_2O. On the other hand, in catalysis by nucleophiles, no such effects operate, and the catalytic rate constants are approximately the same in both H_2O and D_2O.

In the hydrolysis of acetic anhydride, the catalytic rate constant for formate ion is approximately the same in both H_2O and D_2O. Thus the formate ion is a nucleophile catalyst in this reaction, as indicated in Eqs. (3-32) and (3-33). Of course, it is possible that some catalytic events occur with formate ion as a general base; however, the major catalytic pathway must be that of nucleophile catalysis.

Nucleophile-catalyzed reactions, in contrast to general base-catalyzed reactions, involve an unstable intermediate compound. Conversely, evidence that such an intermediate occurs in a reaction constitutes excellent proof that the reaction is nucleophile-catalyzed rather than general base-catalyzed. However, a failure to observe such an intermediate does not rule out nucleophile catalysis. It is possible that the intermediate hydrolyzes more readily than it is formed. In such a case the intermediate is present in a small amount and difficult to detect. This is true for acetic formic anhydride, the intermediate in our example.

Another example of nucleophile catalysis is shown in Eqs. (3-36) and (3-37), in which imidazole acts as a nucleophile catalyst in the hydrolysis of the ester *p*-nitrophenyl acetate.

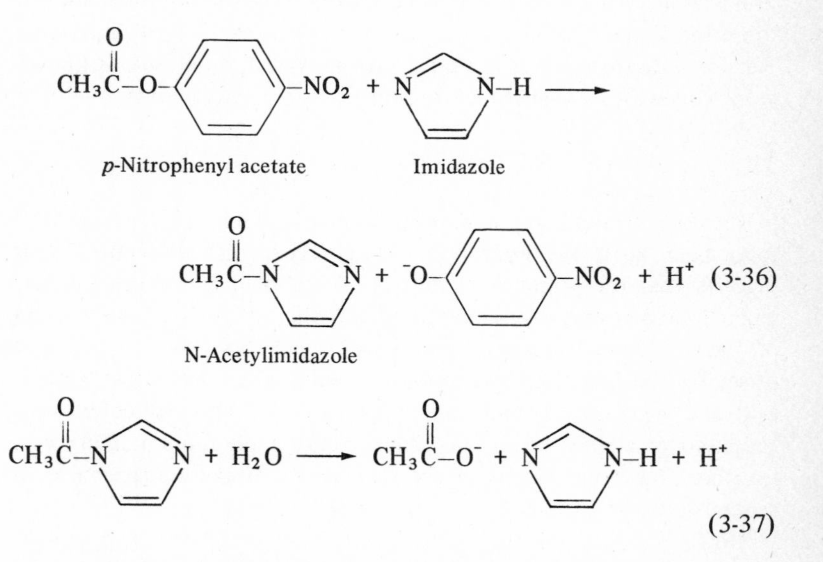

In this case a high concentration of imidazole causes the first step of the reaction to occur faster than the second step. Consequently, the intermediate *N*-acetylimidazole accumulates in the solution, and its presence has been detected. Thus, the mechanism of nucleophile catalysis is established for this reaction. Also, as expected for such a mechanism, the catalytic rate constant for imidazole is the same in D_2O as in H_2O.

We have just spoken of detecting the intermediate *N*-acetylimidazole in a reaction. This compound is a well-known crystalline substance. It is colorless, which means that it does not absorb light in the visible region of the spectrum. However, it does absorb ultraviolet light in the region of wavelength 245 nm. Thus one may conveniently "observe" the appearance and disappearance of this intermediate during the course of the reaction by a spectrophotometric technique. A spectrophotometer is an instrument which passes a beam of light of a selected wavelength through a solution and determines how much of the light is absorbed by the solution. A recording spectrophotometer makes a permanent recording of how the absorbance of the solution changes during the course of the reaction. When the reaction between *p*-nitrophenyl acetate and imidazole was performed in a recording spectrophotometer set at a wavelength of 245 nm, the absorbance was observed first to increase and then decrease corresponding to the appearance and disappearance of *N*-acetylimidazole. In this way the intermediate was "observed." The intermediate was confirmed as *N*-acetylimidazole because it disappeared at the same rate as *N*-acetylimidazole is known to hydrolyze. It was subsequently isolated in crystalline form.

3-9 MULTIPLE CATALYSIS

In previous sections we have noticed that general acids and general bases both assist the hydrolysis of carboxylic acid derivatives. This suggests that the presence of both a general acid and a general base might have a cooperative catalytic effect which is greater than the sum of their individual catalytic effects. Although this may be true, it is offset by the fact that four molecules (substrate, nucleophile, general acid, and general base) have to line up in just the right way before such cooperative catalysis can occur. This is considerably more unlikely than for three molecules to be in the right orientation for reaction as in general acid and general base catalysis individually.

However, reactions are known where a base catalyst can assist a nucleophile catalyst in hydrolysis reactions. An example is the hydrolysis of phenyl acetate shown in Eq. (3-38).

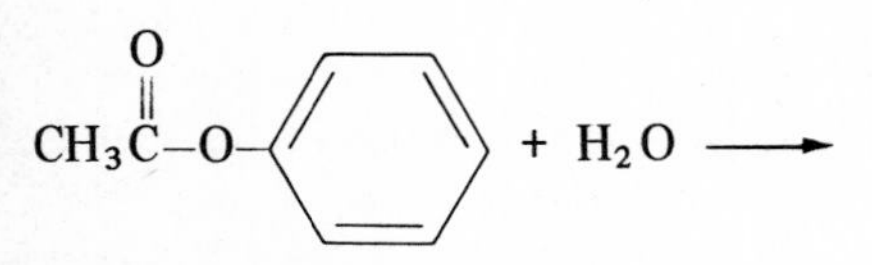

Phenyl acetate

$$CH_3\overset{\overset{\large O}{\|}}{C}{-}O^- + HO{-}C_6H_5 + H^+ \qquad (3\text{-}38)$$

Phenol

Imidazole is a nucleophile catalyst for this reaction, but if the solution is made basic by adding some sodium hydroxide, the OH^- ion can act as a proton acceptor to assist imidazole in attacking the substrate. This is illustrated by the transition-state diagram in Eq. (3-39).

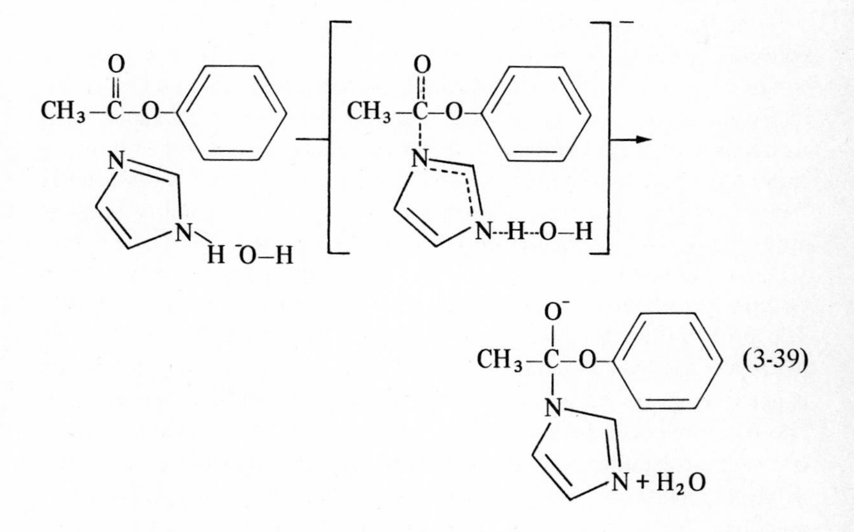

The tetrahedral intermediate produced in this reaction can then proceed to the final products as described in previous sections in this chapter. As an exercise the student is urged to write out all the steps for the remainder of the reaction. Other general bases can assist the reaction in Eq. (3-39) in the same way as the OH^- ion. In fact a second molecule of imidazole can act as the general base, accepting the proton from the nucleophilic imidazole molecule. Thus, in Eq. (3-39) the hydroxide ion is acting as a general base.

3-10 INTRAMOLECULAR CATALYSIS

In previous sections of this chapter, we have noted that general acids, general bases, and nucleophiles can catalyze the hydrolysis of carboxylic acid derivatives. We have "explained" these catalyses in terms of electronic distortions in the reacting molecules in the case of acid-base catalysis and the formation of reactive intermediates in the case of nucleophile catalysis. In general acid and base catalyses, a collision among *three* molecules must occur.

If the three molecules are completely independent of each other, as they would be in the gas phase for example, they would have to collide *simultaneously* in the correct orientation; in this sense, the word simultaneous means over a period of time of less than one or two vibrations, that is, approximately 10^{-13} s.

This is such an extremely unlikely event that even if all such collisions resulted in reaction, the rate of reaction by this pathway would be very low. However, in aqueous solutions, acids and bases are hydrogen-bonded to the substrate and/or water molecules and form relatively long-lived complexes. In general base catalysis, for example, a catalyzed reaction requires the collision of a *preexisting* water-general base complex with the substrate molecule, a probability closely approaching that of a collision between *two* species rather than *three*.

In the example of Sec. 3-6, the catalyzed and uncatalyzed reactions occurred at equal rates for an imidazole concentration of 0.0036 *M*. The probabilities for collision of substrate with water and with water hydrogen-bonded to catalyst are proportional to the corresponding concentrations of 55 and 0.0036 *M*, if one water molecule is bonded to one base molecule. In the catalyzed pathway, this unfavorable ratio of collision probabilities is offset by the fact that the presence of the catalyst facilitates the rearrangement of electrons to give rise to reaction. In general, the collision probability is a factor which contributes to the entropy part, $\Delta S^{\ddagger}$, of the free energy of activation $\Delta G^{\ddagger}$, with $\Delta S^{\ddagger}$ becoming more negative as the probability decreases; the facilitation of electron rearrangement has the effect of decreasing the enthalpy of activation $\Delta H^{\ddagger}$. These two effects tend to cancel according to Eq. (3-40). (See also Sec. 1-2.)

$$\Delta G^{\ddagger} = \Delta H^{\ddagger} - T\Delta S^{\ddagger} \qquad (3\text{-}40)$$

In the example cited, the values of $\Delta G^{\ddagger}$ for the catalyzed and uncatalyzed pathways are the same at a catalyst concentration of 0.0036 *M*, since the rates are the same. However, with catalyst at unit concentration (that is, 1 *M*), which is the usual, or standard, condition for comparison, $\Delta G^{\ddagger}$ for the catalyzed pathway is considerably less than $\Delta G^{\ddagger}$ for the uncatalyzed pathway. This statement is equivalent to saying that at 1-*M* catalyst concentration, the catalyzed pathway in this example predominates greatly over the uncatalyzed pathway.

We may summarize the preceding remarks by recalling that the rate of a catalyzed reaction is dependent on the catalyst concentration, the reason being that as the catalyst concentration is increased, the likelihood of reaction increases because more collisions between catalyst and reactants occur each second. That is, the likelihood of a reaction occurring increases as the probability of collision between catalyst and reactant increases. This leads us to the idea of intramolecular catalysis. If a catalytic group is present within the substrate molecule itself in such a way that it is accessible to the site of reaction, the situation is fundamentally the same as if an external catalyst were present at extremely high concentration. We shall consider three examples of intramolecular catalysis: the hydrolyses of aspirin, of *o*-carboxyphenyl-β-D-glucoside, and of monophenyl succinate.

The hydrolysis of aspirin (acetylsalicylate) is interesting because it was one of the first cases of intramolecular catalysis recognized. *A priori*, at neutral pH the anionic form of the carboxyl group could act either as a nucleophile catalyst [giving an anhydride as an intermediate, (see Sec. 3-8)] or as a general base catalyst. A wide range of evidence, including a 2.2 times decrease in rate in D_2O as compared with H_2O, points toward general base catalysis in the rate-determining step, with the formation of the tetrahedral intermediate, as shown in Eq. (3-41). The subsequent breakdown of the tetrahedral intermediate may involve general acid catalysis, as indicated in Eq. (3-42); however, such catalysis cannot be detected kinetically because this step is not the rate-determining step.

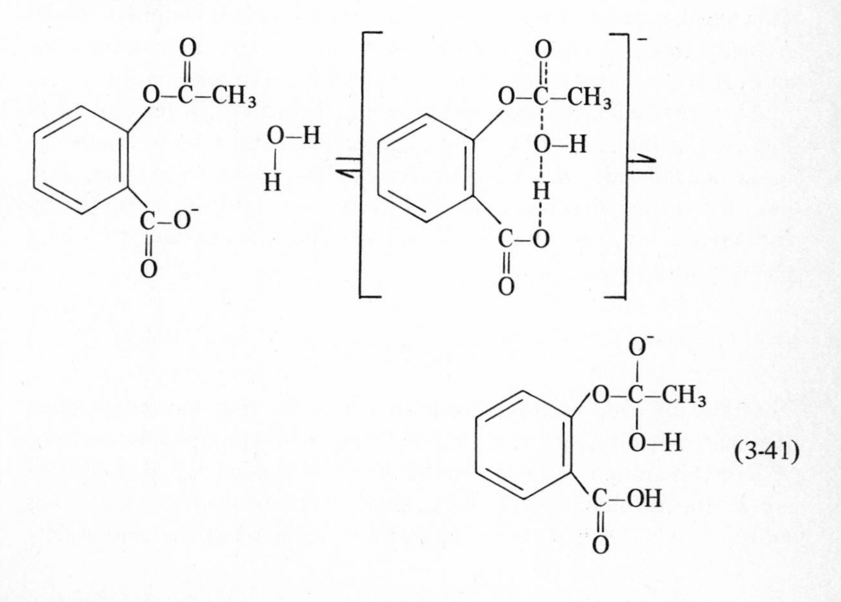

(3-41)

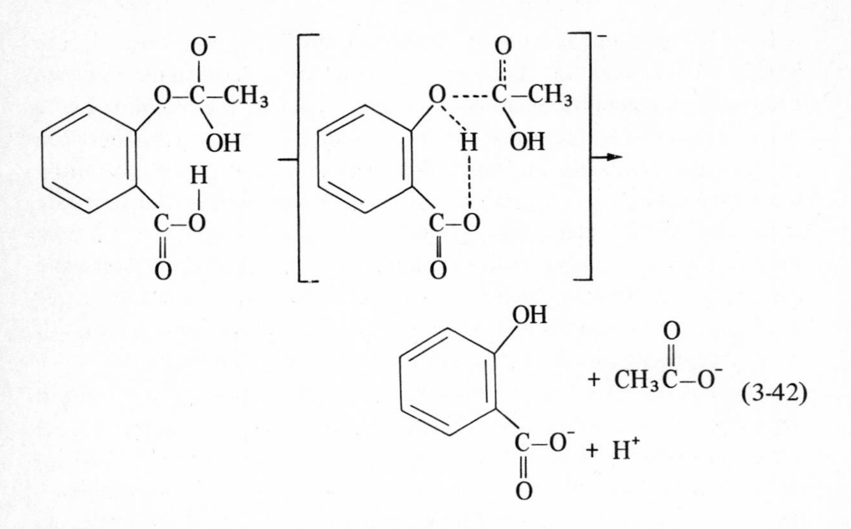

$$\text{(3-42)}$$

The internal catalysis of aspirin hydrolysis can occur only when the carboxyl group is in the anionic or basic form. In acidic solutions where the carboxyl group is protonated, external catalysis by hydronium ion occurs. In sufficiently basic solutions, hydrolysis by direct attack of hydroxide ion on the carbonyl carbon of the ester group becomes more important than internal general base catalysis by carboxylate ion. These types of catalysis are illustrated by Fig. 3-3, a pH-rate profile showing the rate constant for aspirin hydrolysis as a function of pH. Actually this is a log-log plot since k is plotted on a logarithmic scale and pH is the negative logarithm of the hydronium ion concentration.

The observed rate constant for aspirin hydrolysis, k (as plotted in Fig. 3-3), is the sum of three terms according to Eq. (3-43), where k_{H^+} is the second-order rate constant for catalysis by external hydronium ion, k_i is the first-order rate constant for internal catalysis by carboxylate ion, and k_{OH^-} is the second-order rate constant for direct reaction with hydroxide ion.

$$k = k_{H^+}[H_3O^+] + k_i \frac{[RCOO^-]}{[RCOOH] + [RCOO^-]} + k_{OH^-}[OH^-] \qquad (3\text{-}43)$$

The ratio of concentrations associated with the rate constant k_i gives the degree of dissociation of the internal carboxyl group in aspirin. The pK_a of this group is 3.6. Thus, above pH 5.5 almost all aspirin molecules are in the carboxylate ion form, and the ratio of concentrations is unity. At pH 3.6 half of the aspirin molecules are in the carboxylate (basic) form, and so the rate is only one-half the maximum observed between pH 6 and pH 8. At pH 2.6 the ratio of carboxyl to carboxylate groups is 10:1; thus the fraction of carboxylate ions is 0.091. Hence at pH 2.6 the contribution of internal catalysis to the observed rate is only 9.1 percent of its contribution at pH 6 to pH 8. However, at this pH the hydronium ion term also makes a small contribution, so that the observed value of k is the sum of these two terms. Below pH 1.5 the only important term is the hydronium ion term. Above pH 3 the H_3O^+ concentration is too small to contribute significantly to the hydrolysis of aspirin. From pH 3 to pH 8, only internally catalyzed hydrolysis [the second term in Eq. (3-43)] is important. Above pH 10 only the hydroxide ion term is important. At pH values between the regions stated, two terms contribute significantly to the observed rate constant as noted above, for example, for pH 2.6. If the internal catalysis by

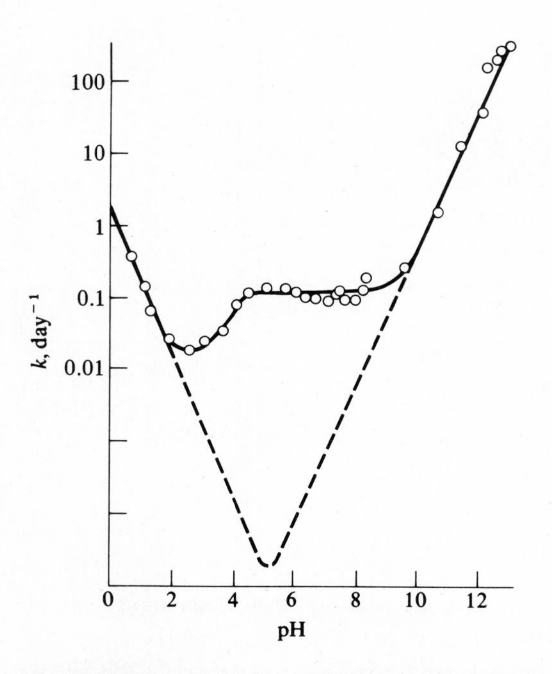

FIGURE 3-3 Variation in the rate of aspirin hydrolysis with pH. [*From M. L. Bender, "Mechanisms of Catalysis of Nucleophilic Reactions of Carboxylic Acid Derivatives," Chemical Review, 60, 53 (1960).]*

carboxylate ion did not occur, the rate profile would take on the shape indicated by the dotted line in Fig. 3-3.

The reader may wish to calculate the half-life of aspirin *in vitro* when it is taken as an analgesic. The pHs of stomach contents, small intestine, and blood are 2, 8, and 7, respectively. Assume that the rate of hydrolysis at 37°C (body temperature) is three times faster than at 25°C (as in Fig. 3-3).

Another example of intramolecular catalysis by a carboxyl group is the hydrolysis of *o*-carboxyphenyl-β-D-glucoside [Eq. (3-44)], which is 10^4 times faster than the hydrolysis of the similar compound having the carboxyl group in the para position. Although the hydrolysis of acetals via external catalysis is hydronium ion-catalyzed (Sec. 3-2), an internal group can act as a general catalyst. This reaction is of interest because in this case the carboxyl group is a general acid catalyst in contrast to the aspirin case, and it serves as a model for the enzyme lysozyme (Sec. 3-13).

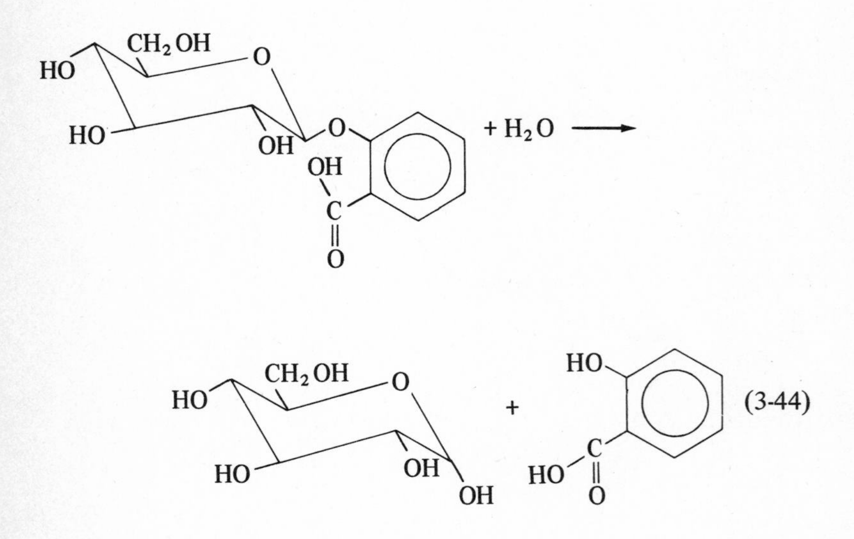

Here the internal carboxyl group donates a proton to the oxygen atom involved in the carbon-oxygen bond that breaks [Eq. (3-44)].

In the hydrolysis of mono-*p*-bromophenyl succinate, the carboxylate group acts as a nucleophile catalyst as shown in Eqs. (3-45) and (3-46), not as a general acid or base catalyst.

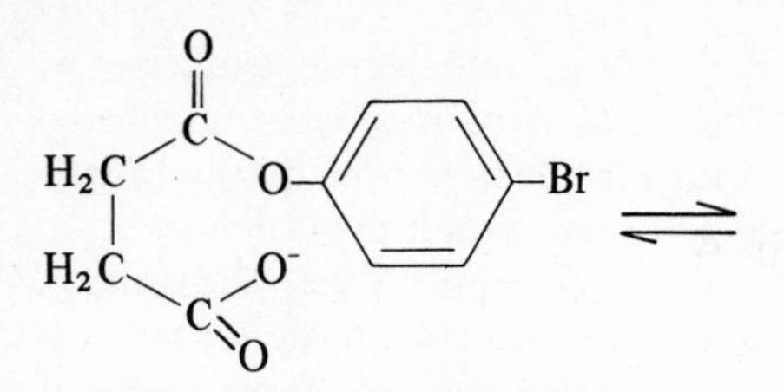

Mono-*p*-bromophenyl succinate

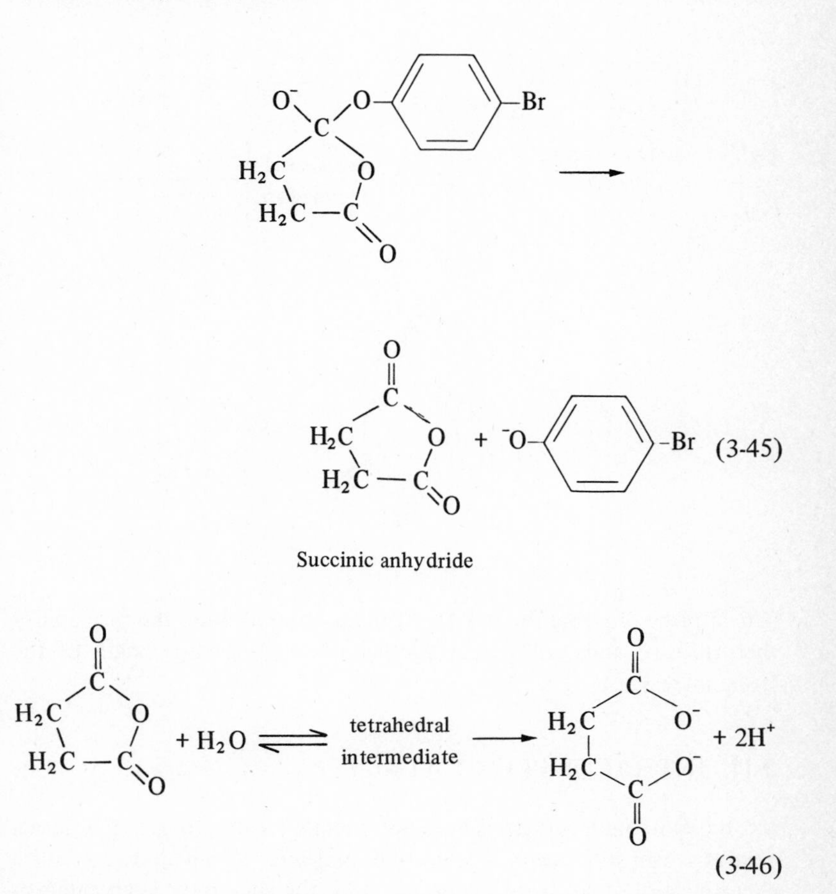

Succinic anhydride

(3-45)

(3-46)

This reaction is analogous to the formate ion-catalyzed hydrolysis of acetic anhydride discussed in Sec. 3-8. In this case the anhydride intermediate (succinic anhydride) is a five-membered ring, a size of ring which is highly favored by the natural tetrahedral geometry of saturated carbon compounds. The reaction of mono-*p*-bromophenyl glutarate to form glutaric anhydride (the first step in the hydrolysis reaction), shown in Eq. (3-47), is 230 times slower than the formation of succinic anhydride from the analogous succinate ester. The reaction of the glutaric compound is slower largely because, being longer, it is less likely to achieve a conformation in which the two reacting parts of the molecule come close together.

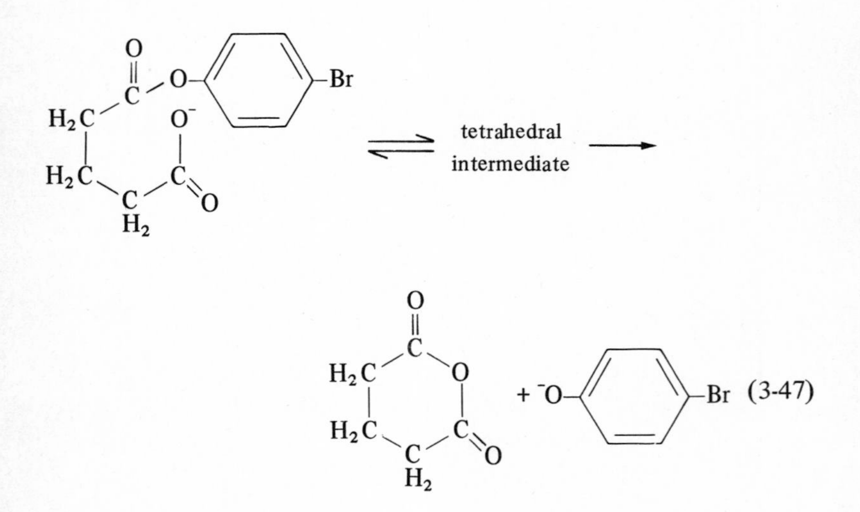

(For a piece of string floating freely in an agitated fluid, the probability that the two ends will come together decreases as the length of the string increases.)

3-11 INTRACOMPLEX CATALYSIS

If a substrate molecule itself does not contain a catalytic group, internal catalysis can still occur if a second molecule which does contain a catalytic group can form a complex with the substrate. Such catalysis may be called intracomplex catalysis.

As a first example we shall consider the hydrolysis of *p*-nitrophenyl acetate as catalyzed by the anion of *o*-mercaptobenzoic acid. The first step in the mechanism for this reaction is shown in Eq. (3-48).

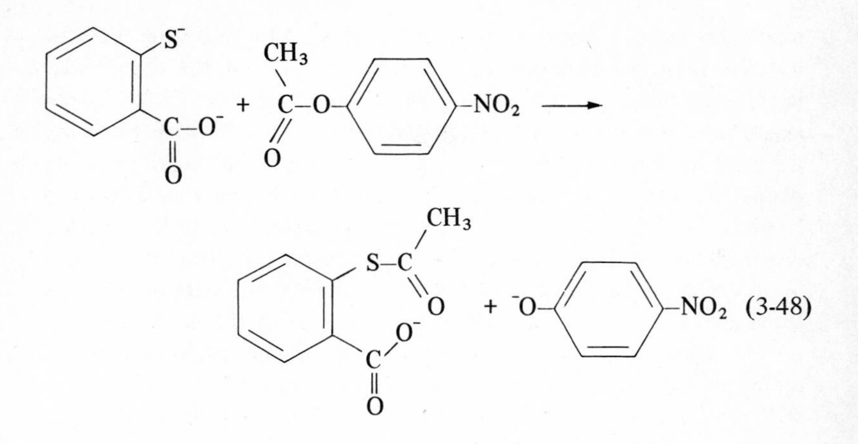

The product of this reaction is called thioaspirin because it is identical to aspirin except that a sulfur atom replaces one of the oxygen atoms of aspirin. Thioaspirin probably hydrolyzes by the same mechanism as aspirin, although it is possible that the carboxylate ion in this case may act as a nucleophile catalyst rather than a general base catalyst. This reaction is an example of intracomplex catalysis because thioaspirin may be considered to be a covalent complex between *p*-nitrophenyl acetate and the *o*-mercaptobenzoate ion. The uncatalyzed hydrolysis of *p*-nitrophenyl acetate proceeds as discussed in Sec. 3-4. In the presence of the catalyst *o*-mercaptobenzoate ion, the hydrolysis is accelerated by two types of catalysis: nucleophile catalysis, the formation of the covalent complex thioaspirin; then intracomplex, general base (or nucleophile) catalysis by the carboxylate ion. The net result is that *p*-nitrophenyl acetate hydrolyzes and the *o*-mercaptobenzoate ion is chemically unchanged.

Our ultimate goal is to understand how enzymes exert their catalytic effect. In the preceding example, *o*-mercaptobenzoate ion is like an enzyme in the sense that it first binds to the substrate and then completes the hydrolysis reaction by the catalytic effect of another of its functional groups. However, it has a very low catalytic efficiency compared to enzyme efficiencies. Furthermore, enzymes generally form noncovalent complexes with substrates, as discussed in Chap. 2, not covalent complexes.

The hydrolysis of *m*-chlorophenyl acetate as catalyzed by cyclohexa-amylose is an example of a reaction in which a noncovalent complex is formed prior to hydrolysis. Cyclohexaamylose consists of six glucose molecules bonded together in such a way that the resulting molecule has the shape of a doughnut. The inner surface of the doughnut is largely nonpolar. Furthermore, the diameter of the "hole" is five angstrom units, which is large enough to accept a substituted phenyl group. (See Fig. 3-4.) Since the *m*-chlorophenyl group of *m*-chlorophenyl acetate is also nonpolar, it can enter the cavity of cyclohexaamylose and form a relatively strong "inclusion" complex which is noncovalent but held together by apolar forces, i.e., the attraction of nonpolar groups for each other. In this complex, the ester group in the substrate comes into close proximity with one of the many hydroxyl groups which line the top and bottom of the cyclohexaamylose molecule, and nucleophile catalysis by the hydroxyl group occurs as shown in Eqs. (3-49) and (3-50). Subsequently, the *m*-chlorophenol

FIGURE 3-4 Cyclohexaamylose with phenylacetate starting to enter the "hole."

molecule dissociates from cyclohexaamylose leaving the latter free to repeat the catalytic cycle with another molecule of substrate.

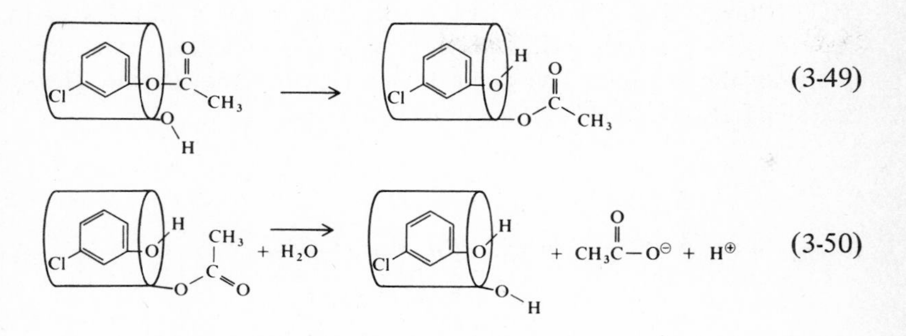

This mechanism is formally analogous to the mechanism for the action of hydrolytic enzymes such as chymotrypsin. We shall now consider these complex biological catalysts.

3-12 ENZYME CATALYSIS: CHYMOTRYPSIN

Chymotrypsin is a digestive enzyme which catalyzes the hydrolysis of ingested proteins in the small intestine. It is also a catalyst for the hydrolysis of amino acid esters and other carboxylic acid derivatives. It shows particular specificity for derivatives of the aromatic amino acids phenylalanine, tyrosine, and tryptophan. We shall consider chymotrypsin as an example of hydrolytic enzymes. It is the most intensively studied of all enzymes and the major aspects of its mechanism of action are now known. Certain other hydrolytic enzymes are believed to operate by the same, or similar, mechanism. Among others, these include: trypsin and elastase, which are digestive proteolytic enzymes like chymotrypsin; acetylcholinesterase, which is found in various tissues in mammals; subtilisin, found in the bacterium *Bacillus subtilis*; and papain, ficin, and bromelain, plant enzymes found in papaya latex, fig latex, and pineapple stem, respectively.

The overall mechanism by which chymotrypsin catalyzes the hydrolysis reactions of carboxylic acid derivatives is shown in Eq. (3-51). The formation of a noncovalent complex between enzyme and substrate, the Michaelis-Menten complex, is characteristic of all enzyme-catalyzed reactions of specific substrates. The existence of such a complex is implied by the phenomenon of enzyme saturation, as discussed in Chap. 2. The mechanism of Eq. (3-51) also involves an acyl-enzyme intermediate (R–C(=O)–O–E) which implies that one type of catalysis involved in the reaction is nucleophile catalysis. The second step, for which a rate constant k_2 is indicated, is called the acylation step since the substrate acyl group is transferred to the enzyme. The final step (rate constant k_3) is the deacylation step.

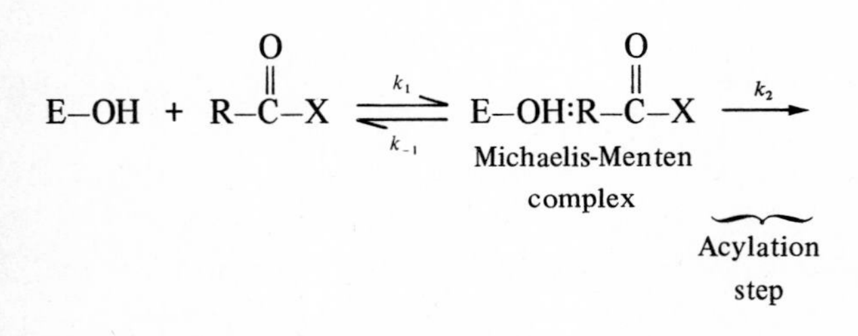

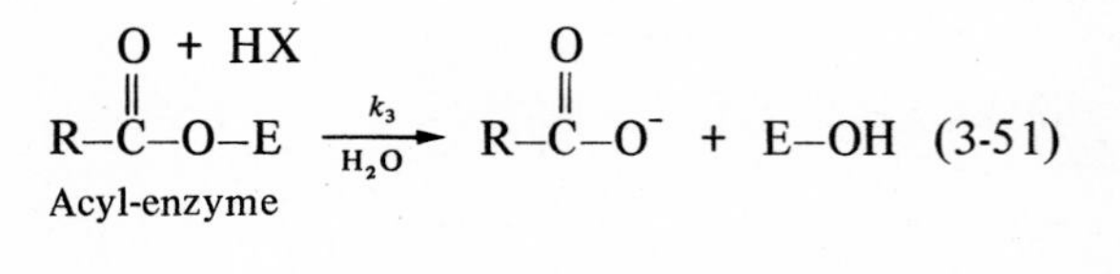

We shall consider first some of the evidence for the existence of an acyl-enzyme intermediate. In 1949 it was discovered that the nerve gas diisopropylphosphorofluoridate reacts with chymotrypsin and makes it catalytically inactive. In this reaction only one molecule of the nerve gas reacts with each molecule of chymotrypsin. The inactive enzyme was subsequently hydrolyzed to its constituent amino acids by heating it in 6 *M* HCl at 110° for one day. It was found that the diisopropylphosphoryl group was attached to the amino acid serine. Since there are 29 serine residues in chymotrypsin and only one reacts with the nerve gas, the one which reacts must be special in some way and is probably at the active site. In this case the substrate diisopropylphosphorofluoridate reacts with chymotrypsin according to Eq. (3-51), except that $k_3 = 0$ (the acyl-enzyme is stable). Hence, the substrate is not a specific one for chymotrypsin, since the overall hydrolysis reaction [Eq. (3-52)] does not go to completion.

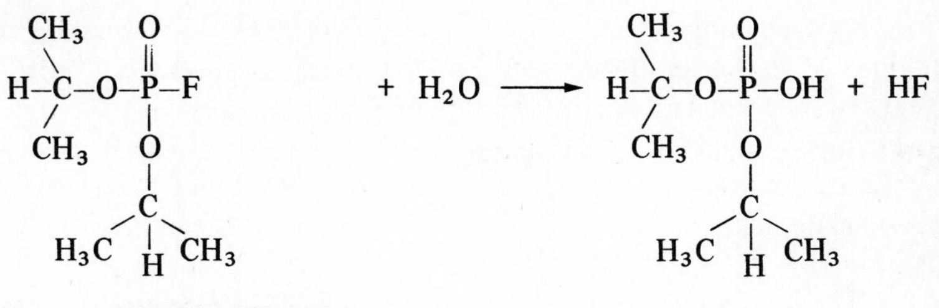

Diisopropylphosphorofluoridate

(3-52)

The question arises whether specific substrates also give an acyl-enzyme intermediate. The hydrolyses of such substrates, by definition, are very efficiently catalyzed and an acyl-enzyme, if it exists, will have a very fleeting existence and be extremely difficult to detect. In spite of this, an acyl-enzyme has been observed in the chymotrypsin-catalyzed hydrolysis of the methyl ester of *N*-furylacryloyl-L-tryptophan in the millisecond time range using a special instrument called a stopped-flow spectrophotometer.

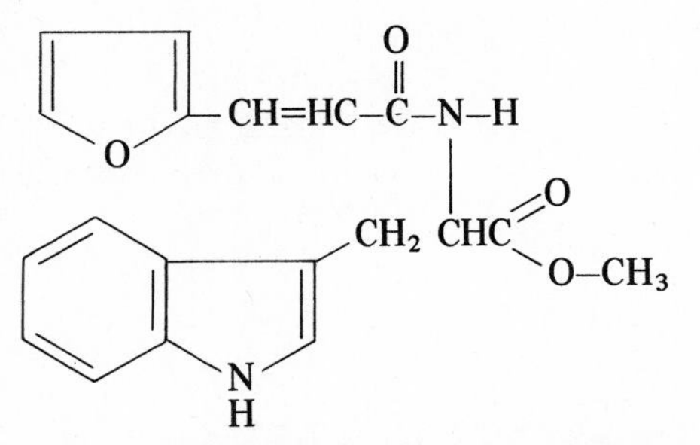

Methyl ester of *N*-furylacryloyl-L-tryptophan

An indirect approach has also been used to demonstrate the existence of an acyl-enzyme in chymotrypsin-catalyzed reactions. If the substrate is such that the rate constant k_2 in Eq. (3-51) is much greater than k_3, and if the enzyme is saturated with substrate [i.e., $[S]_0 \gg K_s$, (see Sec. 2-3)], then almost all of the enzyme is rapidly converted into acyl-enzyme. At the same time there will be released an amount of HX equal to the amount of enzyme in solution. At a slower rate, determined by k_3, acyl-enzyme reacts with water to regenerate free enzyme which immediately reacts with more substrate to give acyl-enzyme and more HX. Thus the rate of production of HX, observed from the beginning of the reaction, is as shown in Fig. 3-5. Initially, an amount of HX is produced very rapidly, commonly called a "burst," followed by a slower and constant rate of HX production.

Figure 3-5 illustrates only about the first 10 percent of a typical reaction. The part of the reaction following the burst is called the steady-state part because the concentrations of intermediates during this part of the reaction remain essentially constant or "steady." The burst part of the reaction, called the pre-steady-state, may be complete within a few milliseconds with specific substrates and thus may be difficult to observe directly. However, even if the pre-steady-state is not directly observed, the existence of a burst can be inferred if the steady-state part of the curve in Fig. 3-5 extrapolates back to intersect the ordinate ($t = 0$) at a value of $[HX] > 0$. If such a burst occurs in a reaction, it constitutes evidence for an acyl-enzyme intermediate. A very useful additional piece of information from a burst experiment is that the value of the intercept in Fig. 3-5 is equal to the concentration of enzyme in the solution. This is a method for titrating the enzyme. Convenient substrates for this type of experiment are *p*-nitrophenyl esters of specific substrates because HX in Eq. (3-51) is then *p*-nitrophenol (or its anion) which absorbs light very strongly. Thus the reaction may be readily followed using a spectrophotometer.

Based on the preceding experiments, it is clear that a group at the active site of chymotrypsin acts as a nucleophile catalyst. What is

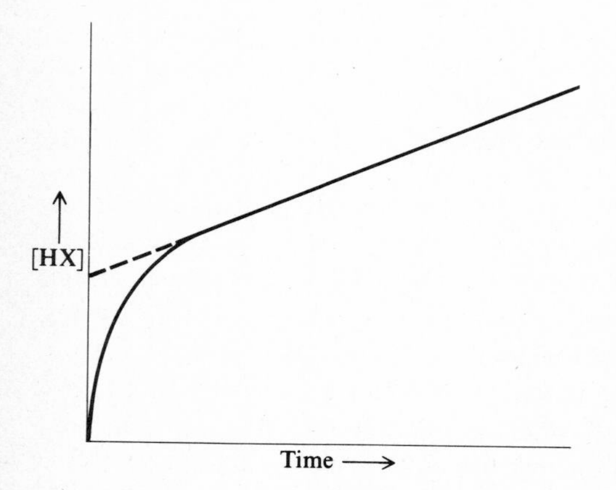

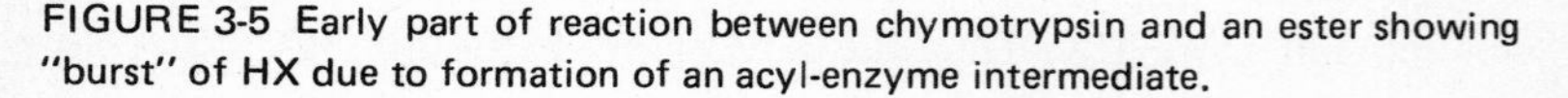

FIGURE 3-5 Early part of reaction between chymotrypsin and an ester showing "burst" of HX due to formation of an acyl-enzyme intermediate.

the identity of this group? The experiment with diisopropylphosphorofluoridate suggests that it is the hydroxyl group of a serine residue. But diisopropylphosphorofluoridate is not a good substrate. Does it actually react at the active site? An indication that it does is that the reaction is inhibited in the presence of 3-phenylpropanoic acid. Since the latter compound is a competitive inhibitor in reactions of specific substrates, it must bind at the active site; the fact that it inhibits the reaction of diisopropylphosphorofluoridate with chymotrypsin implies that diisopropylphosphorofluoridate also reacts at the active site. Therefore, these experiments indicate that the hydroxyl group of serine is the nucleophile catalyst at the active site.

The following experiment provides further confirmation that a serine hydroxyl group participates in the catalytic process. It is possible to dehydrate a serine residue in chymotrypsin to give "anhydrochymotrypsin" [Eq. (3-53)]. This modified enzyme is inactive.

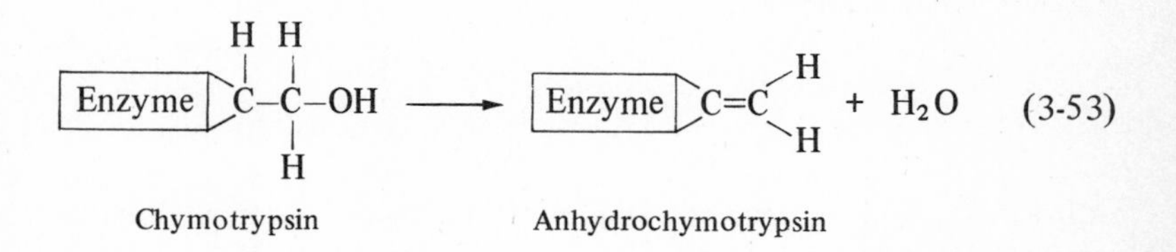

The serine residue affected is the same one which is involved in the reaction with diisopropylphosphorofluoridate. The loss of activity associated with dehydration does not prove that the serine is part of the active site. It is possible that the serine is distant from the active site and that dehydration causes the enzyme to take up a different conformation which is inactive. However, it was further shown that anhydrochymotrypsin is still capable of forming Michaelis-Menten complexes with specific substrates even though no hydrolysis occurs. Thus, the active site must remain relatively intact when the serine residue is dehydrated. Hence the reason for loss of catalytic activity must be that the serine hydroxyl group plays a direct role in the catalytic action of chymotrypsin.

Besides the serine hydroxyl group, certain other functional groups in chymotrypsin which are acids and bases are involved in the catalytic process. This is evident from pH-rate profiles such as that shown in Fig. 3-6 for the hydrolysis of *N*-acetyl-L-tryptophanamide. The

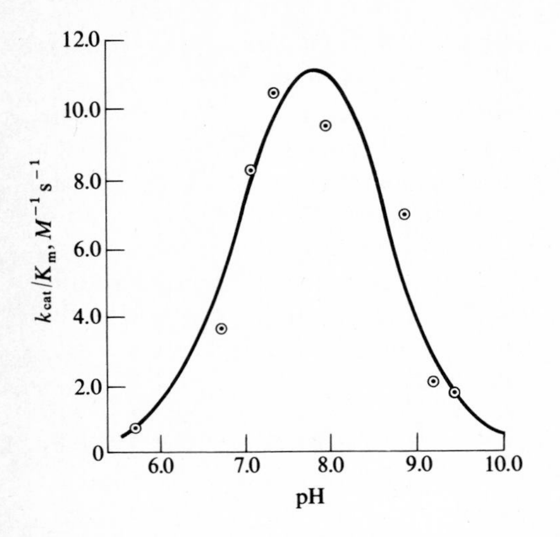

FIGURE 3-6 pH-rate profile for the chymotrypsin-catalyzed hydrolysis of *N*-acetyl-L-tryptophanamide.

ordinate k_{cat}/K_m is generally the most readily determined kinetic parameter in enzyme kinetics. These quantities were discussed in Sec. 2-4. The bell-shaped profile of the points in Fig. 3-6 implies that two ionizable groups with pK_a's of approximately 7 and 9 affect the activity of chymotrypsin. The group of pK_a 7 must be in its dissociated, or basic, form, and the group of pK_a 9 must be in its undissociated, or acidic, form for the enzyme to be fully active. The curve in Fig. 3-6 is the theoretical profile for dependence of the reaction on two such ionizable groups. The deviations of the points from the theoretical line are within acceptable experimental error. Of the ionizable groups known to be present in isolated amino acid side chains, only the imidazole group of histidine has a pK_a of 7. Thus the presence of a histidine residue at the active site is implied, but no means proved. For example, the hydroxyl group of a tyrosine, which has a pK_a of 10 when free tyrosine is dissolved in water, might have a pK_a of 7 in the active site due to environmental influences.

Additional evidence is available that the imidazole group in histidine is a part of the active site. Chymotrypsin is inactivated in the presence of the chloroketone, 1-chloro-4-phenyl-3-toluenesulfonamido-2-butanone.

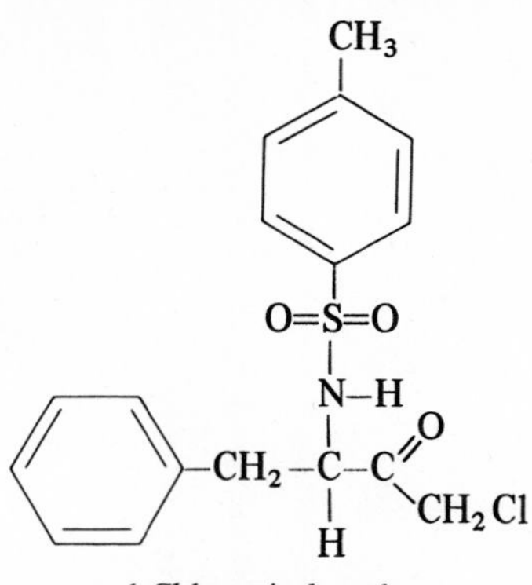

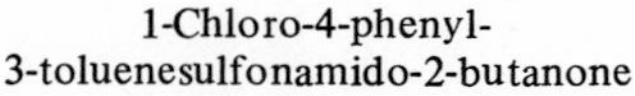
1-Chloro-4-phenyl-
3-toluenesulfonamido-2-butanone

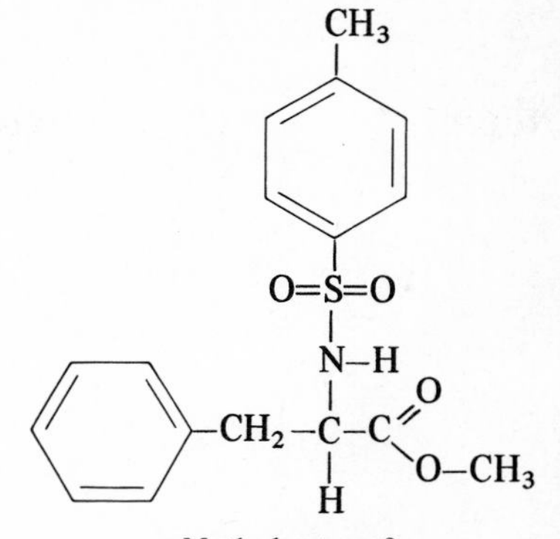

Methyl ester of
N-toluenesulfonyl-L-phenylalanine

Subsequent analysis of the inactive enzyme shows that only one ketone molecule reacts with each enzyme molecule, and that the ketone reacts with the imidazole group of one particular histidine residue. The reaction is a nucleophilic displacement of chlorine in the ketone by imidazole, similar to the reaction in Sec. 1-1. This particular ketone was designed to be structurally similar to specific substrates, such as derivatives of *N*-toluenesulfonyl-L-phenylalanine, so it would bind to the active site. Thus, the imidazole group with which it reacts is undoubtedly located at the active site and not elsewhere.

What is the function of the imidazole group in chymotrypsin? *A priori*, it may function either as a nucleophile catalyst or as a general base catalyst. Comparative kinetic studies in H_2O and in D_2O indicate that it acts as a general base catalyst. In reactions for which $k_2 \gg k_3$, the observed catalytic rate constant k_{cat} becomes equal to k_3 [see Eq. (2-21)]. Thus a study of factors affecting k_{cat} is really a study of factors affecting k_3, the deacylation rate constant. A typical substrate of chymotrypsin for which $k_2 \gg k_3$ is the ethyl ester of *N*-acetyl-L-tryptophan. The values of k_{cat} as a function of pH for this substrate are shown in Fig. 3-7; the lines drawn through the experimental points are calculated assuming that the reaction is catalyzed by a base (imidazole) of pK_a 6.9 in water, and a base of pK_a 7.2 in D_2O. Thus in H_2O at pH 6.9 only half of the imidazole catalytic groups are in the basic form, and the observed rate is only half of the maximum rate observed at higher pHs where all the imidazole is in the basic form. In D_2O the increase in pK_a is to be expected; the catalytic group is still an imidazole group in the enzyme. The rates in D_2O and H_2O can be

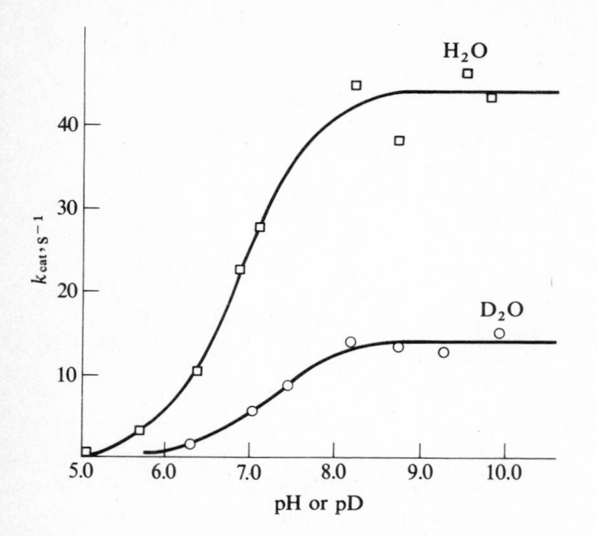

FIGURE 3-7 The variation in rate constant k_{cat} with pH for the chymotrypsin-catalyzed hydrolysis of the ethyl ester of *N*-acetyl-L-tryptophan. [*From M. L. Bender et al., "pH Dependence of α-Chymotrypsin-Catalyzed Reactions," Journal of the American Chemical Society, 86, 3686 (1964).*]

legitimately compared only when the fraction of the imidazole in the basic form is the same in both solvents. This is most conveniently done above pH 9 where all the imidazole is in the basic form in both solvents. Since the rate is 2.7 times faster in H_2O than in D_2O, we can conclude that the imidazole group is a general base catalyst, not a nucleophile catalyst. When the hydrogen atom on the nitrogen atom of imidazole is replaced by a methyl group, the enzyme is catalytically inactive. This indicates that the imidazole acts as a general acid catalyst as well as a general base catalyst.

With other substrates the acylation rate constant k_2 shows the same dependence on pH as illustrated for k_3 in Fig. 3-7 and the same sensitivity to whether the solvent is H_2O or D_2O. Thus the acylation step is also assisted by general base catalysis by imidazole.

We may now write down the mechanism of chymotrypsin-catalyzed reactions showing specifically how a serine hydroxyl group and a histidine imidazole participate in the reaction. This is given in Eq. (3-54), where the bulk of the enzyme is illustrated schematically by a line joining the two important catalytic groups.

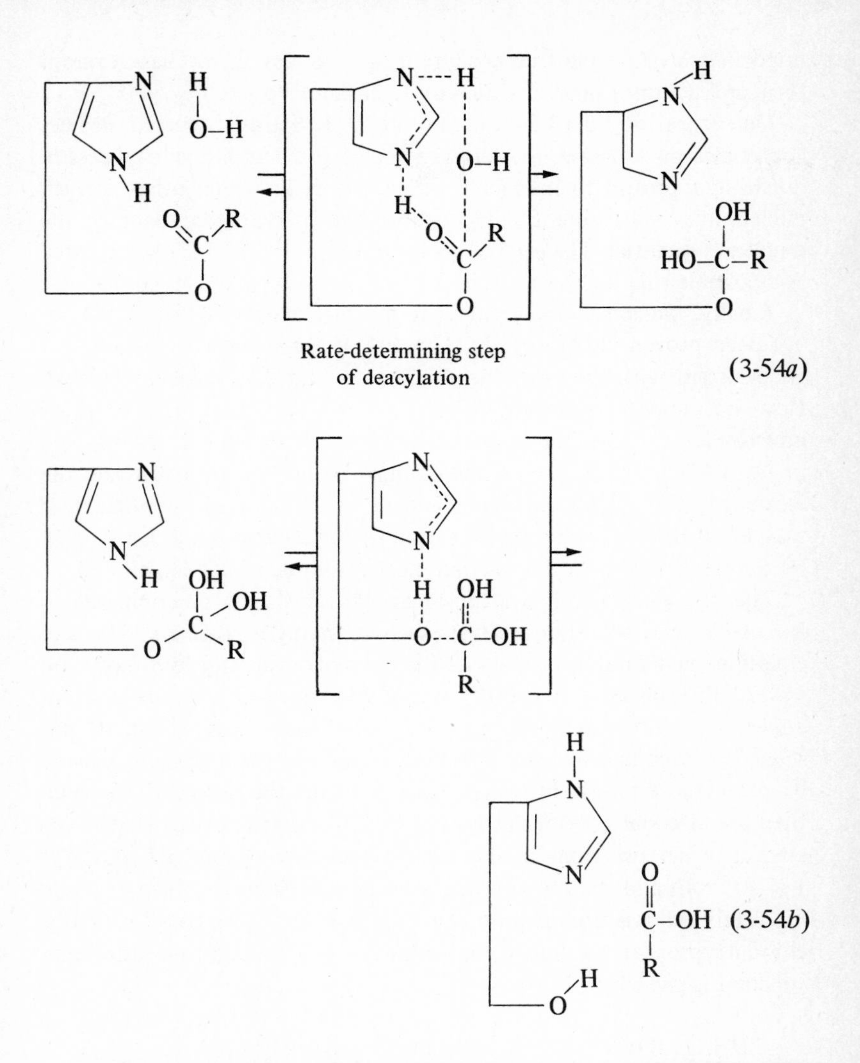

To avoid unnecessary repetition, only the diagram for the deacylation step is shown. The acylation step is exactly the reverse except that the carboxylic acid is replaced by a derivative thereof, and the water molecule is replaced by the leaving group of the carboxylic acid derivative, e.g., ethanol in the case of an ethyl ester substrate. The final diagram in Eq. (3-54*b*) may be considered a Michaelis-Menten complex between enzyme and product which would be in rapid equilibrium with free enzyme and free product. Such a complex is generally not important because the free product acid is usually in the basic (anion) form and does not bind significantly to chymotrypsin.

One aspect of Eq. (3-54*a*) has not been previously discussed. Besides acting as a nucleophile, the water molecule also acts as a general acid in donating a proton to imidazole. Experiments in which other oxygen nucleophiles substitute for H_2O show that a hydrogen atom on the oxygen is essential. The general acid catalysis in Eq. (3-54*a*) is suggested as a possible role for this proton.

Finally, we must emphasize that the mechanism of Eq. (3-54) has not been proven unequivocally. The test of a mechanism is whether it is consistent with the experimental data, and Eq. (3-54) meets this test. However, some recent experiments indicate that there may be more intermediates in chymotrypsin-catalyzed reactions than the two shown in Eq. (3-51). Further experiments must be devised to investigate the existence of these intermediates and to determine their identity. This may eventually require a modification of the mechanism in Eq. (3-54). However, the main features of that mechanism seem well established.

The efficiency of chymotrypsin as a catalyst may be compared to the efficiencies of other catalysts such as the hydroxide ion. The rate equations and catalytic constants for chymotrypsin and hydroxide ion toward the substrate *N*-acetyl-L-tryptophanamide (ATrA) are given in Table 3-2. Unfortunately, k_{cat} is a first-order rate constant and k_{OH^-} is a second-order rate constant; hence the two constants cannot be compared directly. Instead we shall compare the two catalysts when they are at equal concentrations at pH 8, the approximate pH at which chymotrypsin functions in the digestive process. At this pH, $[OH^-] = 1 \times 10^{-6}\ M$, and the pseudo-first-order rate constant for the hydroxide ion-catalyzed reaction is given by $k_{OH^-}[OH^-] = 3 \times 10^{-10}\ s^{-1}$. For chymotrypsin at the same concentration the pseudo-first-order rate constant is given by

$$\frac{k_{cat}\,[E]}{K_m + [S]} = \frac{0.057\ s^{-1} \times 1 \times 10^{-6}\ M}{0.0052\,M + 0.001\,M} = 9 \times 10^{-6}\ s^{-1}$$

where we have arbitrarily assumed that the substrate concentration is $1 \times 10^{-3}\ M$. Under these conditions chymotrypsin is clearly a better catalyst by a factor of 30,000. Even this number is misleadingly small since the actual concentration of chymotrypsin at pH 8 can be made considerably larger than $1 \times 10^{-6}\ M$, whereas the concentration of hydroxide ion is fixed by the pH.

TABLE 3-2 Comparison of two catalysts for the hydrolysis of *N*-acetyl-L-tryptophenamide

Chymotrypsin + ATrA (pH 8)	$\text{Rate} = \frac{k_{cat}[E]}{K_m + [S]}[S]$	$k_{cat} = 0.057\ s^{-1}$	$K_m = 0.0052\ M$
OH^- + ATrA	$\text{Rate} = k_{OH^-}[OH^-][S]$	$k_{OH^-} = 3 \times 10^{-4}\ M^{-1} \cdot s^{-1}$	

Why is chymotrypsin such an efficient catalyst? Qualitatively, the reason is that the active site of the enzyme combines several types of catalytic groups which can operate simultaneously on the substrate. As shown in Eq. (3-54), general base catalysis, general acid catalysis, and nucleophile catalysis all operate simultaneously. Furthermore, a specific substrate is apparently bound to the active site in such a way that the part of the substrate which undergoes reaction is favorably oriented with respect to these catalytic groups. Thus chymotrypsin is simply a catalyst of the intracomplex type in which three types of catalysis operate after binding.

3-13 ENZYME CATALYSIS: LYSOZYME

Lysozyme is a hydrolytic enzyme of about 14,600 molecular weight. Each molecule is a single polypeptide chain of 129 amino acids. The sequence of amino acids in the molecule is shown in Fig. 3-8. Lysozyme became suddenly popular in 1965 when Professor D. C. Phillips and his colleagues at the Royal Institution in London reported the three-dimensional structure of lysozyme as determined by x-ray analysis. The position of each of the atoms in the molecule is known to within 0.25 Å, which is quite small. Lysozyme is the first enzyme for which such a detailed three-dimensional structure has been determined, and a knowledge of its structure has been exceedingly helpful in determining its mechanism of action, as we shall see.

Lysozyme is a protective enzyme. It is found, among other places, in nasal mucus, sputum, and tears. By promoting the dissolution of bacterial cell walls, lysozyme protects against infection. It does so by catalyzing the hydrolysis of the polysaccharide which constitutes part of the bacterial cell wall. This polysaccharide is a long polymer containing alternating glycoside residues of *N*-acetyl-β-glucosamine and *N*-acetyl-β-muramic acid joined by β-(1-4) glycosidic bonds. Such bonds

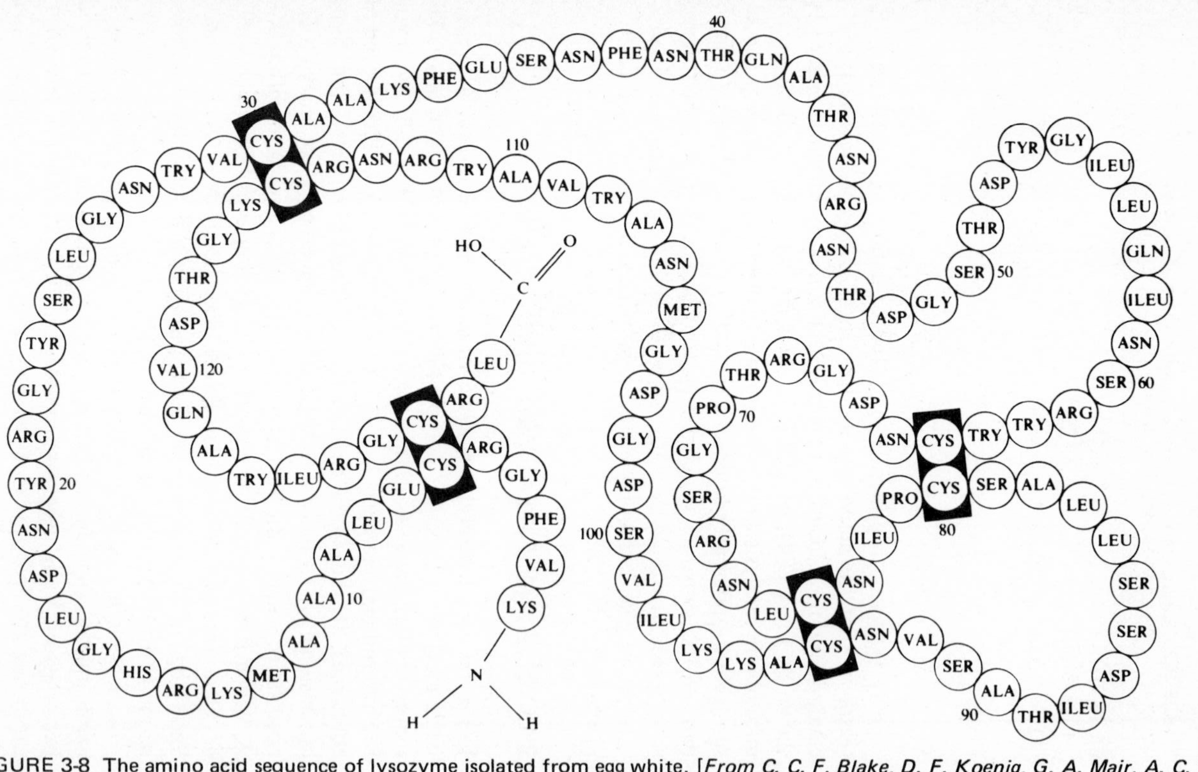

FIGURE 3-8 The amino acid sequence of lysozyme isolated from egg white. [*From C. C. F. Blake, D. F. Koenig, G. A. Mair, A. C. T. North, D. C. Phillips, and V. R. Sarma, "Structure of Hen Egg-White Lysozyme," Nature, 206, 757 (1965)*.]

fall into the category of acetal bonds. (See Sec. 3-2.) When carbon-1 of a glycoside ring is not bonded to another glycosidic residue, it is in equilibrium with an open-chain form in which carbon-1 is part of an aldehyde group as shown in Eq. (3-55). The open-chain form is frequently called the reducing form because aldehydes are reducing agents. Either of two ring forms can result upon cyclization of the open form: the β species with the hydroxyl group bonded to carbon-1 on the same side of the ring as the carbon-6 tail; or the α species with the hydroxyl group on carbon-1 on the opposite side.

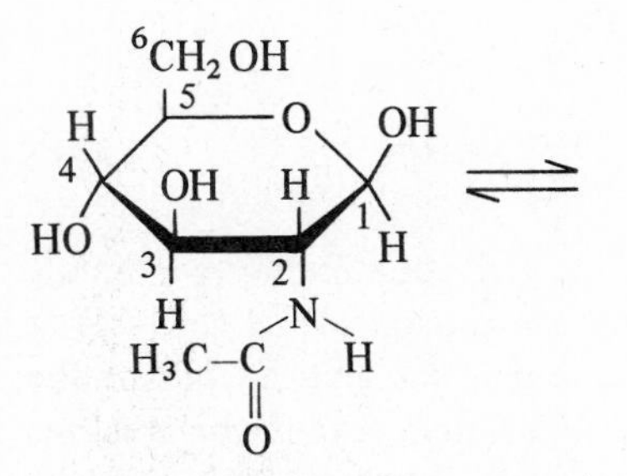

N-Acetyl-β-glucosamine

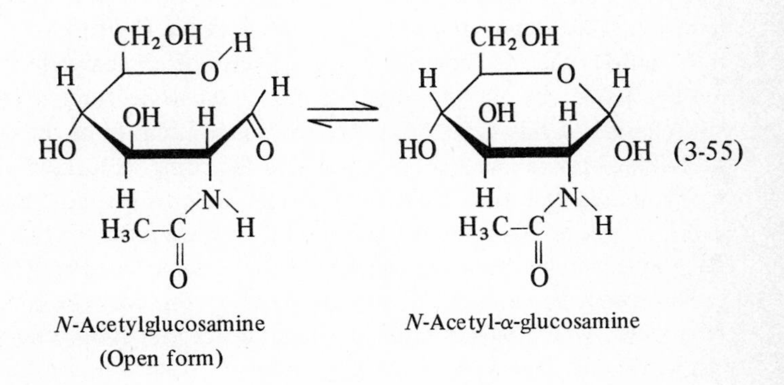

N-Acetylglucosamine
(Open form)

N-Acetyl-α-glucosamine

To illustrate the structure of the polysaccharide, a four-sugar segment, a tetrasaccharide, is shown in Eq. (3-56).

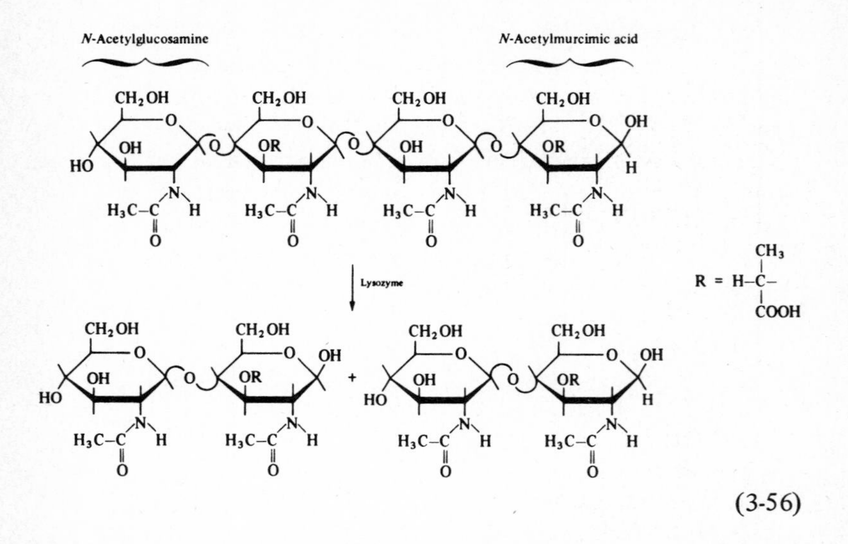

(3-56)

N-Acetylmuramic acid differs from *N*-acetylglucosamine by containing an additional group indicated by R. By comparison with Eq. (3-55), it is clear that the right-hand end is the reducing end and can exist in the α configuration as well as the β configuration. Lysozyme exhibits specificity in that it catalyzes the hydrolysis of a bond such that the new reducing end which forms belongs to an *N*-acetylmuramic acid residue. Thus, lysozyme catalyzes the hydrolysis of only one of the three glycosidic bonds in the tetrasaccharide in Eq. (3-56).

A polysaccharide consisting exclusively of *N*-acetylglucosamine residues joined by β-(1-4) bonds is also a substrate for lysozyme. A trisaccharide of this type, however, acts as an inhibitor of the enzyme. By making an x-ray analysis of crystals of the complex formed between lysozyme and this trisaccharide, it was possible to pinpoint with high precision where the inhibitor is bound to the enzyme. It is located in the groove which runs down one side of the enzyme (see Fig. 2-2).

The trisaccharide of *N*-acetyl-β-glucosamine fills only the top half of the groove, which suggests that as many as six glycoside residues in a polysaccharide could bind to this groove. Reasoning in this way, Phillips has built up a hypothetical model of the complex between the enzyme and the hexasaccharide of *N*-acetylglucosamine. This is shown in Fig. 3-9. The six residues are labeled A through F from top to bottom. The reducing end of the substrate is at the bottom, position F. Carbon and oxygen atoms in the substrate are shown as open and filled circles, respectively; the bonds between them are black lines, tapered so

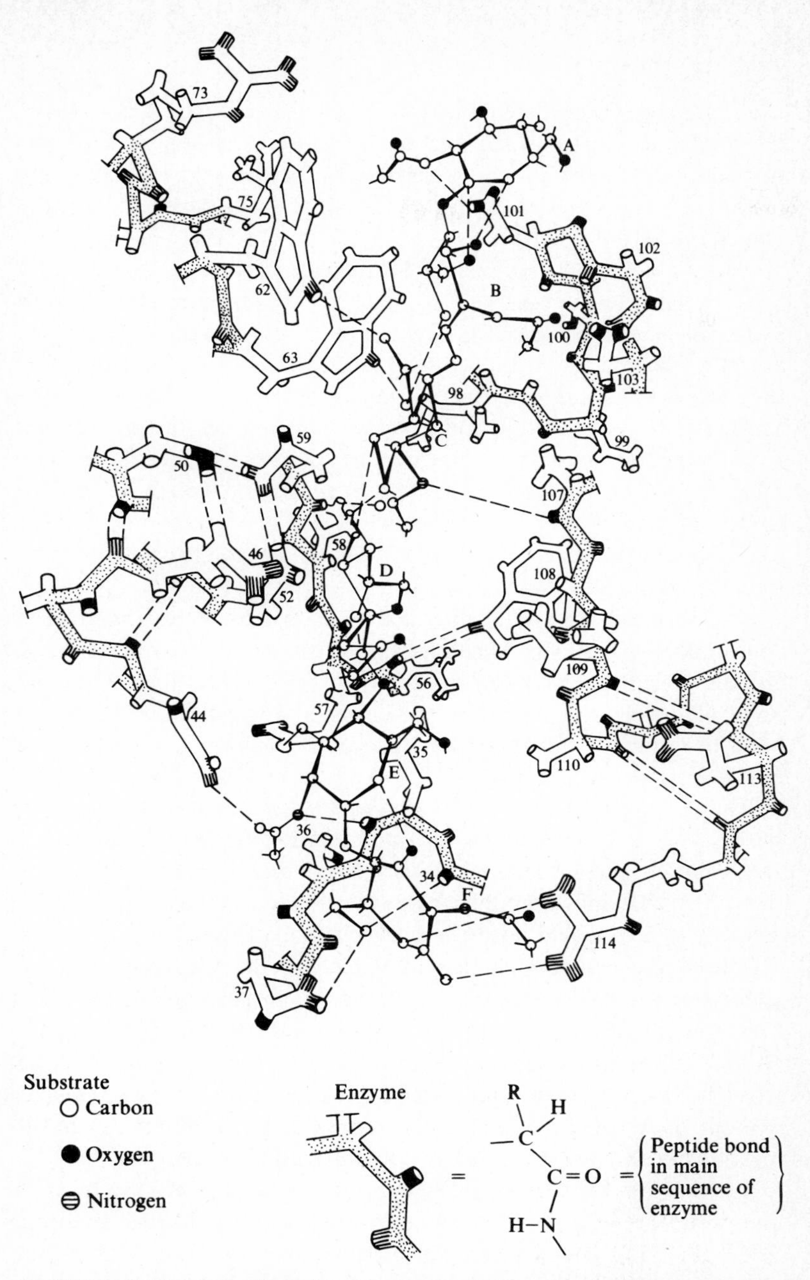

FIGURE 3-9 The three-dimensional structure of the active site of lysozyme. The hexasaccharide of *N*-acetylglucosamine, a substrate, is shown in a possible binding conformation. [*From D. C. Phillips, "The Hen Egg-White Lysozyme Molecule," Proceedings of the National Academy of Sciences, U.S., 57, 484 (1967).*]

as to give the proper three-dimensional perspective. A complete code of the atoms in substrate and enzyme is given in the figure. Only the pertinent part of the enzyme is shown. The amino acid residues in the enzyme are numbered and may be identified by comparison with the sequence in Fig. 3-8.

The single dashed lines in Fig. 3-9 indicate possible hydrogen bonds between substrate and enzyme atoms. The double dashed lines indicate hydrogen bonds between different amino acid residues within the enzyme. The substrate residues A, B, and C in Fig. 3-9 are oriented in the same way as the trisaccharide inhibitor as known from the x-ray analysis. The residues D, E, and F were then fitted into the groove so as to give the maximum possible number of hydrogen bonds between substrate and enzyme. Since the bottom surface of the groove consists largely of nonpolar amino acid residues, apolar interactions of these residues with nonpolar regions of the substrate can also contribute to the total binding force.

The six-atom ring of a glycoside is known to prefer the "chair" conformation. This conformation can be seen most clearly in residue C and also B, and it has been preserved for all residues except D. One point of the chair in residue D, the carbon atom involved in the β-(1-4) bond to residue E, had to be flattened out in order to allow a good fit into the groove.

Thus a reasonable picture of the binding of the hexasaccharide to the enzyme is achieved. The next question is: which bond is cleaved, and what groups on the enzyme assist in the cleavage reaction? Phillips first noted that an *N*-acetylmuramic acid residue with its bulky R group [see Eq. (3-56)] could not fit into the position of residue C in Fig. 3-9. Therefore the natural substrate must bind to the groove so that the alternating *N*-acetylmuramic acid residues are in positions B, D, and F. Consequently, the bond that is hydrolyzed must be the one between B and C or between D and E as implied by Eq. (3-56). But the trisaccharide of *N*-acetylglucosamine binds at positions A, B, and C without undergoing hydrolysis. Therefore, the bond between D and E must be the one that is catalytically hydrolyzed by the enzyme.

In the region between D and E, only two likely prospects for reactive groups are present on the enzyme: amino acids number 35, glutamic acid (glu 35), and number 52, aspartic acid (asp 52). From our discussion in Sec. 3-2 of the hydrolysis of the simple disaccharide, sucrose, we know that acetal hydrolysis is subject to specific hydronium ion catalysis. At pH 6, the normal operating pH of the enzyme, the hydronium ion concentration is too small for lysozyme to utilize this type of specific catalysis. On the basis of the example discussed in Sec. 3-10 [Eq. (3-44)], it is reasonable to suppose that an acid group on the enzyme might participate in general acid catalysis. In fact, Phillips observed that glutamic acid 35 is ideally located for this purpose. It is correctly situated at the bottom of the groove to transfer a proton to the oxygen atom in the acetal bond. Furthermore, it is in a nonpolar environment so that its pK_a should be higher than normal and it should remain in the acidic form up to relatively high pH values. Thus a reasonable first step in the hydrolysis would be a general acid-catalyzed cleavage of the carbon-oxygen bond as shown in Eq. (3-57).

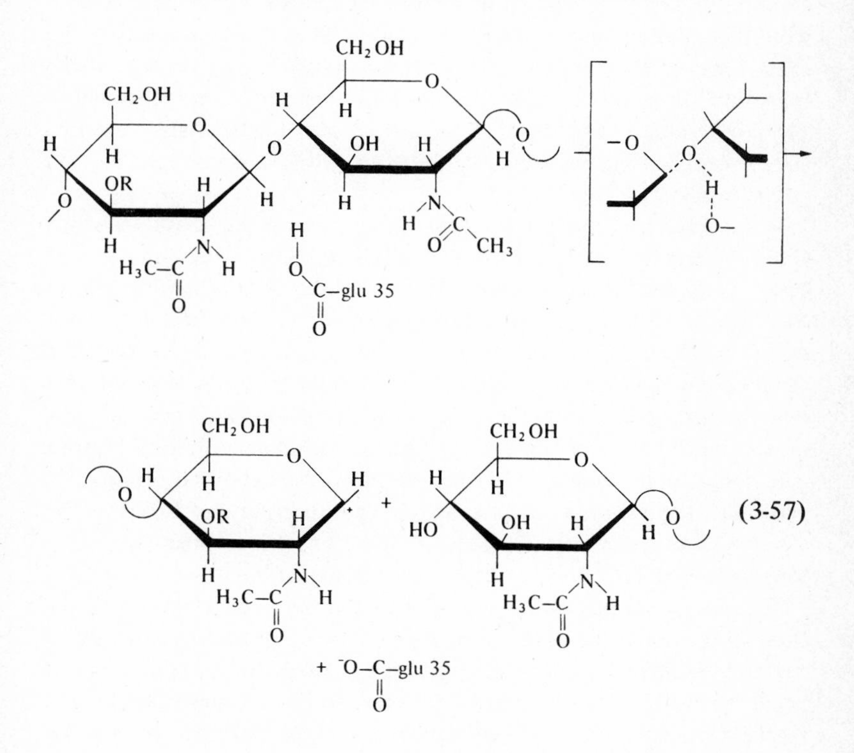

The reaction in Eq. (3-57) is the slow step in the total reaction, since the reaction of water with the product carbonium ion is very fast. Thus, in addition to the general acid catalysis, any factors which can lower the free energy of activation in Eq. (3-57) will increase the efficiency of lysozyme as a catalyst. Part of the free energy of activation is the energy required to create a positive charge on the carbon atom. Phillips noted that the negatively charged carboxylate ion of asp 52 is ideally situated to partially neutralize this positive charge, thereby lowering $\Delta G^{\ddagger}$. Aspartic acid 52 is in a polar environment. Hence its carboxyl group should have a pK_a of 4 or less, and it will be negatively charged at the normal operating pH of 6 for lysozyme.

The carbonium ion product in Eq. (3-57) is most stable in a partially flattened chair conformation. In this conformation the carbon-1 atom can share its positive charge with the ring oxygen by gaining a larger share of the electron pair bonding the two atoms together. For the reaction of Eq. (3-57) to occur in the absence of enzyme, part of $\Delta G^{\ddagger}$ would be the energy required to change the conformation from the chair form of the ground state to the partially flattened form of the transition state. In the proposed enzyme-substrate complex shown in Fig. 3-9 residue D has already been forced into this partially flattened conformation. Thus, binding, of itself, may make a major contribution to the catalytic process.

Thus, Phillips proposed a mechanism for lysozyme action in which three factors contribute to the catalytic process: (1) general acid catalysis by the carboxyl group of glutamic acid 35; (2) stabilization of the transition state through neutralization of the developing positive charge by the carboxylate group of aspartic acid 52; and (3) distortion of the bound substrate. Compared to the uncatalyzed reaction, the first two factors lower the free energy of the transition state and the third factor raises the free energy of the initial state. All these factors contribute to decreasing $\Delta G^{\ddagger}$ compared to the uncatalyzed reaction. In addition, the binding process assists the reaction by bringing the substrate and the catalytic groups on the enzyme into close proximity. This has the effect of raising $\Delta S^{\ddagger}$ and lowering $\Delta G^{\ddagger}$.

The foregoing mechanism was deduced from a consideration of the three-dimensional structure of lysozyme and the known mechanism of acetal hydrolysis. As such it represents a working hypothesis which needs to be tested. A variety of experiments are suggested by the mechanism and some have been carried out. The proposed mechanism is in good accord with three of these experimental results which we shall mention here.

The most obvious prediction of the proposed mechanism is that the hexasaccharide of *N*-acetylglucosamine should be hydrolyzed between the fourth and fifth residue from the nonreducing end. Indeed, when the experiment was performed, two hydrolysis products were obtained, a tetrasaccharide and a disaccharide.

A second experimental observation is that the hexasaccharide of *N*-acetylglucosamine is held to the enzyme less strongly than the trisaccharide. This is consistent with the proposed binding model in Fig. 3-9 if one assumes that the distortion of residue D requires more energy than is regained by the interactions of residues D, E, and F with the enzyme. One could not have predicted this because the strengths of the hydrogen bonds and apolar bonds between enzyme and substrate are not known. However, the result is reasonable.

Finally, experiments show that the carboxyl group of glutamic acid 35 has a pK_a of 6.3. This is considerably higher than the usual value of pK_a 4.0 to 4.5 for carboxyl groups, but it is consistent with the x-ray structure which shows that glutamic acid 35 is in a nonpolar environment. Experiments also show that this acidic group must be in the acid form for the enzyme to be active, which is consistent with the proposed mechanism of general acid catalysis by glutamic acid 35.

The mechanisms of action for chymotrypsin and lysozyme have been studied from two different approaches. What we know about the chymotrypsin mechanism has been learned by chemical studies. The mechanism of lysozyme, on the other hand, has been proposed on the basis of a knowledge of the three-dimensional x-ray structure and a limited amount of chemical information. The proposed mechanism can then be tested by chemical studies. The latter approach, which combines structural and chemical knowledge, is clearly a powerful and efficient one. It is being used more and more as new x-ray analyses become available for other enzymes.

3-14 HETEROGENEOUS CATALYSIS

Thus far in this chapter we have considered only homogeneous catalysis—catalysis in which all reactants and the catalyst are in the same phase. Within this category we have limited ourselves to reactions in aqueous solution. In heterogeneous catalysis, at least two phases are present in the reaction mixture. The most common type is a system with a solid catalyst in contact with substrates in the gaseous state.

Heterogeneous catalysis is similar to enzyme catalysis in that (1) a substrate molecule collides with an active site on the surface of the solid catalyst to form an adsorptive complex, (2) adsorbed substrate reacts in one or more steps under the influence of catalytic groups at the active site, and (3) the product molecules desorb (or escape) from the active site. Thus the concepts of an active site and of complex formation between substrate and active site are common both to enzyme catalysis and heterogeneous catalysis.

A typical particle of the kind of solid catalyst we shall discuss below commonly has a diameter of about 0.05 mm and a large surface area due to its porous nature. Many thousands of catalytic sites may be present on the surface of such a particle. There seem to be several different types of active sites, each type having a different catalytic efficiency. In contrast, a pure enzyme preparation consists of identical protein molecules containing active sites which are identical in structure and catalytic activity. The variability in the active sites of solid heterogeneous catalysts has made mechanistic studies difficult. Nevertheless, heterogeneous catalysts are exceedingly useful and important in our industrial society. As an example of this type of catalysis, we shall consider the catalytic cracking process used in the petroleum industry.

In the past 30 years silica-alumina catalysts have been developed which catalyze the breakdown, or cracking, of large hydrocarbon molecules to smaller hydrocarbon molecules. By this procedure, the high boiling fraction of crude oil can be converted into gasoline and other useful hydrocarbons. Silica-alumina catalysts are oxides in which silicon and aluminum atoms are bonded to each other via oxygen atoms. The whole solid particle, of whatever size, may be viewed as a giant molecule. The presence of some water in the catalyst particle gives rise to acidic groups on the surface, and these are believed to act as general acid catalysts.

In the catalytic cracking process the high-boiling petroleum fraction in the gaseous state (at 500°C) is passed upward through a bed of finely divided solid catalyst. The gas velocity is such that the small catalyst particles become suspended and behave as a liquid. Gaseous hydrocarbons are in the presence of the catalyst only a few seconds. During this time a host of reactions can occur which include reactions not only of the original molecules but of their products as well. We shall consider one possible reaction shown in Eq. (3-58). Step 1 is the formation of the adsorptive complex between substrate and catalyst surface. The nature of the attractive forces involved is not known.

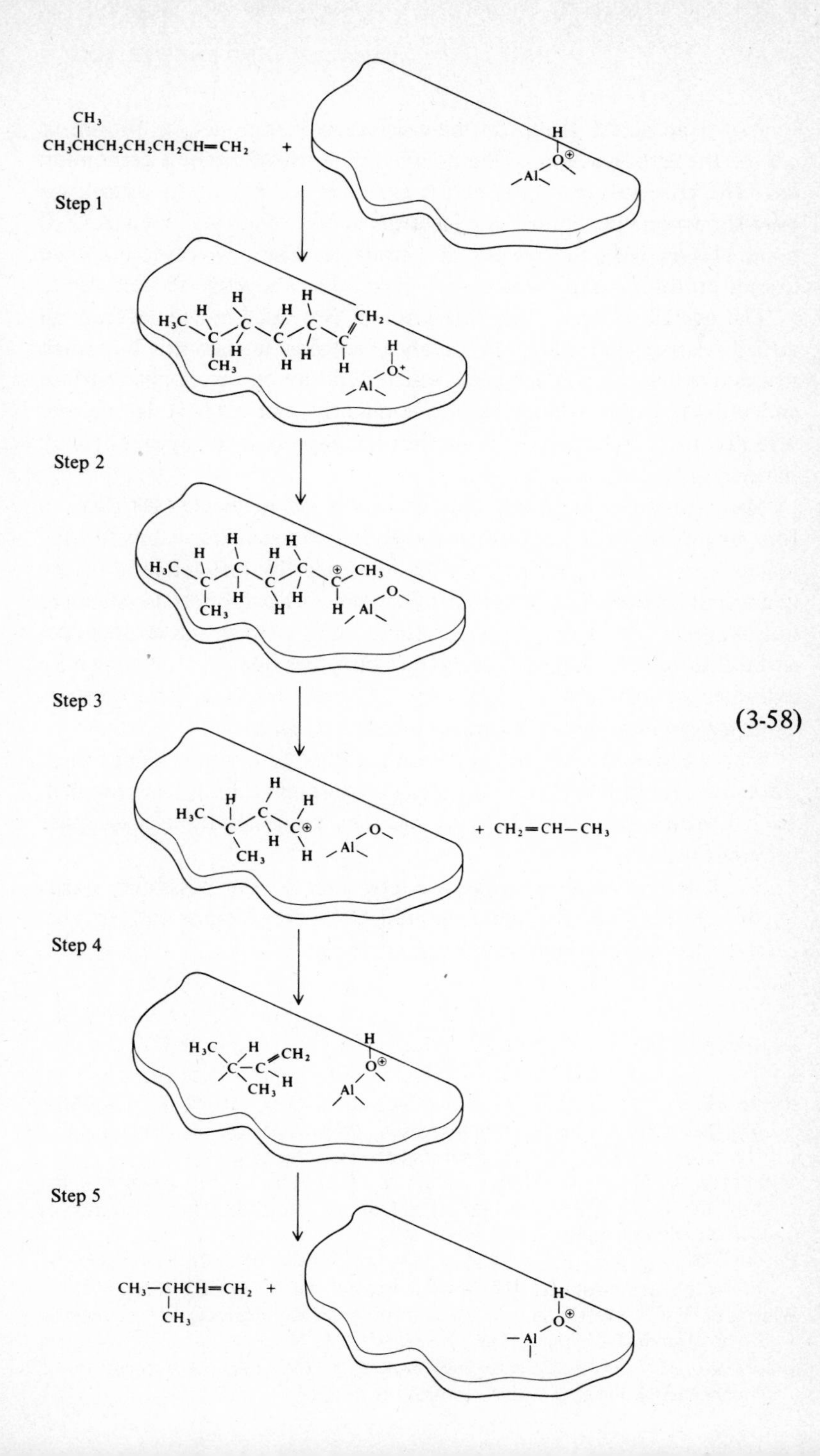

CH₃
CH₃CHCH₂CH₂CH₂CH=CH₂
+
Step 1
Step 2
Step 3
+ CH₂=CH—CH₃
Step 4
Step 5
CH₃—CHCH=CH₂ +
CH₃

(3-58)

In step 2 an acidic group on the catalyst surface donates a proton to one of the carbon atoms of the double bond, thus forming a carbonium ion. The true nature of the acidic group is not definitely known; we have shown one possibility. Step 3 is the actual "cracking" step. A C–C bond cleaves with the release of 1-propene. The new carbonium ion loses a proton in step 4 and desorbs from the catalyst surface in step 5.

The overall process is the formation of two small molecules from an initially larger molecule. The catalyst remains unchanged, but some reaction processes produce coke which deposits on the catalyst surface and quickly makes it inactive. Consequently, the catalyst is removed several times each day, in a continuous process, and regenerated by burning off the coke.

Many enzymes have only one active site per molecule (see Chap. 6 for cases of multiple active sites), whereas heterogeneous catalysts have many active sites per particle. In enzymes all the active sites are necessarily identical, whereas in heterogeneous catalysts the sites are not identical. For example, in the latter one can have active sites on a surface or on an edge of a crystal. Enzymes may be poisoned by extremes of pH and by heavy metal ions, whereas heterogeneous catalysts can be poisoned by traces of sulfur or lead.

Since enzymes and heterogeneous catalysts operate under such different conditions, their catalytic rate constants cannot be compared. Each is efficient in its own way, and they are generally used in quite different reactions.

Solid derivatives of enzymes are now known. The possibility exists in the future that the areas in which heterogeneous and enzyme catalysts are used will overlap more and more.

SUGGESTED READINGS

Bender, M. L.: *Journal of the American Chemical Society*, vol. 73, p. 1626, 1951. This paper reports on the demonstration, by radioisotope tracer methods, of the existence of the tetrahedral intermediate in ester hydrolyses.

Strumeyer, D. H., W. N. White, and D. E. Koshland, Jr.: *Proceedings of the National Academy of Sciences, U.S.*, vol. 50, p. 931, 1963. The preparation of anhydrochymotrypsin is described.

Hartley, B. S., and B. A. Kilby: The Inhibition of α-Chymotrypsin by *p*-Nitrophenyl Phosphate, *Biochemical Journal*, vol. 50, p. 672, 1952.

Whitmore, F. C.: The Common Basis of Intramolecular Rearrangements, *Journal of the American Chemical Society*, vol. 54, p. 3274, 1932.

Heinemann, H.: *Chemical Technology*, vol. 1, p. 286, 1971. A comparison of homogeneous and heterogeneous catalysis is made.

Phillips, D. C.: The Three-dimensional Structure of an Enzyme Molecule, *Scientific American,* p. 78, November, 1966. The scientist who determined the structure of lysozyme by x-ray crystallography discusses the mechanism of this enzyme.

Bender, M. L., F. J. Kezdy, and F. C. Wedler: Chymotrypsin: Enzyme Concentration and Kinetics, *Journal of Chemical Education,* vol. 44, p. 84, 1967. The kinetic equations are derived for the pre-steady-state part of chymotrypsin-catalyzed reactions. Experiments are outlined for using these relationships to titrate chymotrypsin and determine kinetic parameters of substrates.

FOUR
METAL ION CATALYSIS

In the preceding chapter we showed how several elementary types of catalysis, including acid and base catalysis, may be combined to give a considerably more efficient catalyst—an enzyme. In the present chapter we shall consider several types of reactions that are catalyzed by metal ions in solution. Examples have been chosen which are believed to have their counterparts in reactions catalyzed by enzymes which contain a metal ion as part of the active site. In many of these cases the mechanisms for the enzymatic reactions have not been established and may be different from the model systems.

Before looking at metal-ion-catalyzed reactions we must briefly discuss bonding in metal ion complexes and factors which contribute to the stability of such complexes.

4-1 TRANSITION-METAL COMPLEX IONS

The common transition metals are those in which the *d* orbitals are only partially filled by electrons. In solution the positively charged ions of these metals readily combine with negative ions or other small molecules called ligands to form complex ions. A familiar example is the ferricyanide ion in which the six ligands are cyanide ions. The geometry of the ligand arrangement in this case is octahedral. Complex ions with other geometries, such as tetrahedral, square planar, and trigonal bipyramidal, also occur (Fig. 4-1).

Two questions have traditionally aroused the interest of inorganic chemists concerning these complexes: (1) what is the nature of the ligand-metal bond and (2) what determines the geometry of the complex. We do not need to go deeply into this subject. It is sufficient for our purpose to state briefly the current view.

Consider a metal ion, such as the ferric ion, which is capable of forming an octahedral complex with the ligands lying along the x, y, and z axes. The electrons in the ferric ion consist of an argonlike core

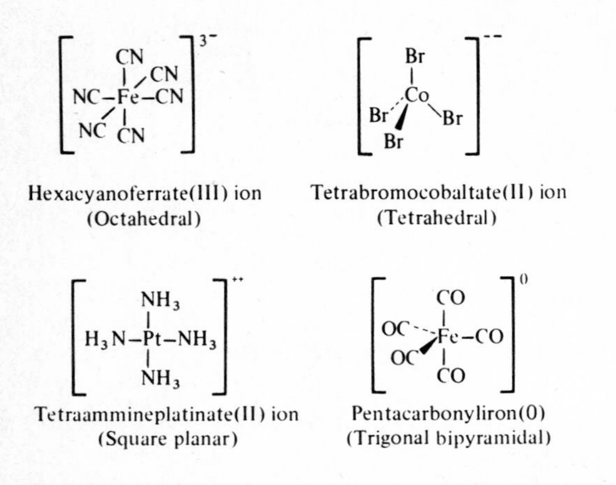

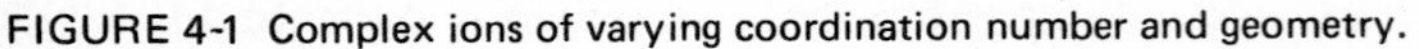
FIGURE 4-1 Complex ions of varying coordination number and geometry.

plus five valence electrons in the five $3d$ orbitals. The $4s$ and $4p$ orbitals lie at just slightly higher energy than the $3d$ level. Thus, a total of nine orbitals of similar energy level are available for filling with electrons. (Their shapes are illustrated in Fig. 4-2.) Six of them, $4s$, $4p_x$, $4p_y$, $4p_z$, $3d_{x^2-y^2}$, $3d_{z^2}$, have lobes lying along the axes towards the ligand positions. When a complex forms, these six orbitals hybridize (mix) with each other (and possibly to a small extent with other orbitals) to give six new orbitals, each with a major lobe pointing in one direction along the axis. The electron pair donated by each ligand overlaps with one of these vacant orbitals giving a covalent bond, called a σ (sigma) bond, which we shall represent pictorially by the usual convention of a straight line. The strength of this bond will vary greatly depending on the nature of the ligand.

The remaining three metal orbitals, $3d_{xy}$, $3d_{yz}$, $3d_{zx}$, point between the ligands and may contain up to six electrons, two in each orbital. (Metal complexes in which the metal ion has more than six d electrons are more difficult to discuss in an elementary fashion, and we shall not attempt to do so.) For example, the ferricyanide complex will have its five valence electrons in these orbitals. Electrons in these orbitals can sometimes participate in what is called π bonding because they will overlap with vacant p or d orbitals in the ligand, if such are present. This is the case in the ferricyanide ion. Thus the iron-cyanide bond

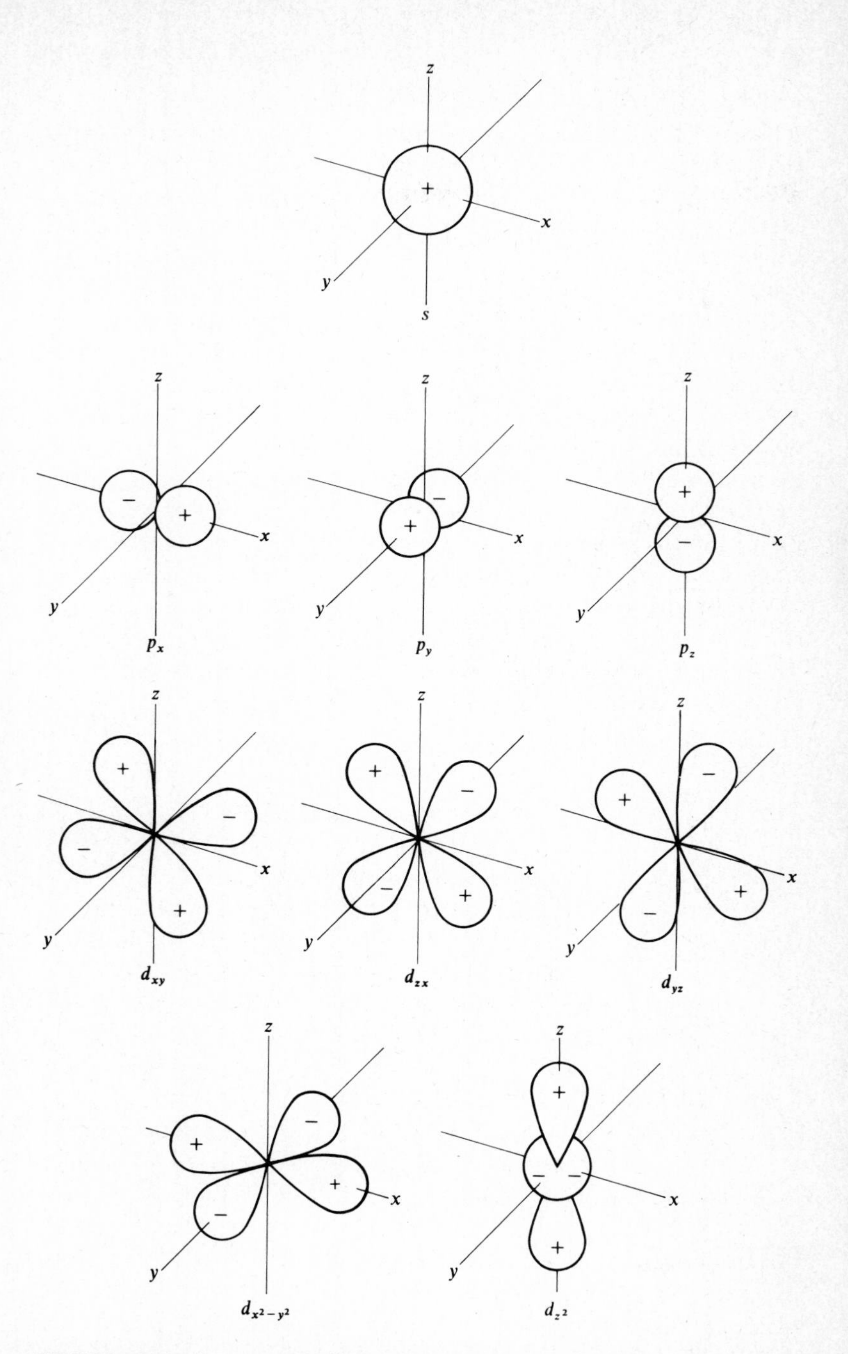

FIGURE 4-2 Atomic orbitals. (*From F. Basolo and R. G. Pearson, "Mechanisms of Inorganic Reactions," 2d ed.*, John Wiley & Sons, Inc., New York, 1967, *p. 54*.)

consists of a σ covalent bond involving an electron pair donated by the ligand and a small amount of π bonding (Fig. 4-3) involving electrons donated by the metal ion. This partial–double-bond character helps to explain why the ferricyanide ion is such a stable ion. For simplicity of nomenclature, such a bond is represented by a single line, but the reader should remember that it has some double-bond character.

The alkenes and alkynes form a class of ligands which we shall encounter later in this chapter. They form both σ bonds and π bonds to the metal ion. The structure of the platinum-ethylene complex in Fig. 4-4*a* has been determined by x-ray analysis which clearly indicates that the metal ion is not bonded to just one carbon atom, but to both of them equally. This is called a three-center bond. Apparently the π electrons from the ethylene double bond overlap one of the vacant metal orbitals to form a σ bond, and one of the electron-containing *d*-orbitals of the metal ion overlaps a vacant *p*-type orbital of ethylene (Fig. 4-4*b*) to form a weak π bond. A complex of this type is often called a π complex. Again, for simplicity we shall represent such a bond by a single line as in Fig. 4-4*a*, while remembering that it has partial–double-bond character.

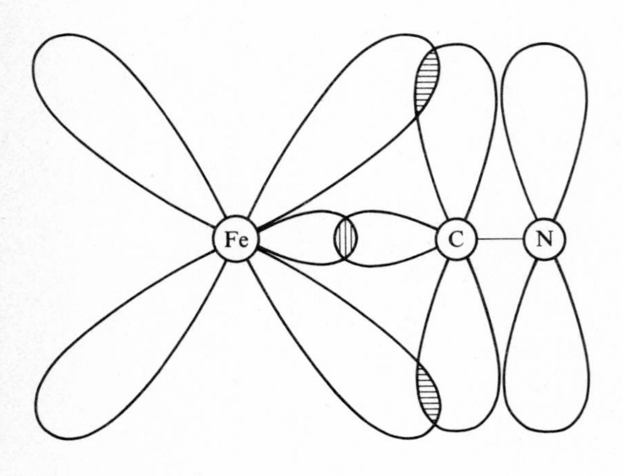

FIGURE 4-3 Schematic representation of iron-cyanide bond in ferricyanide ion.

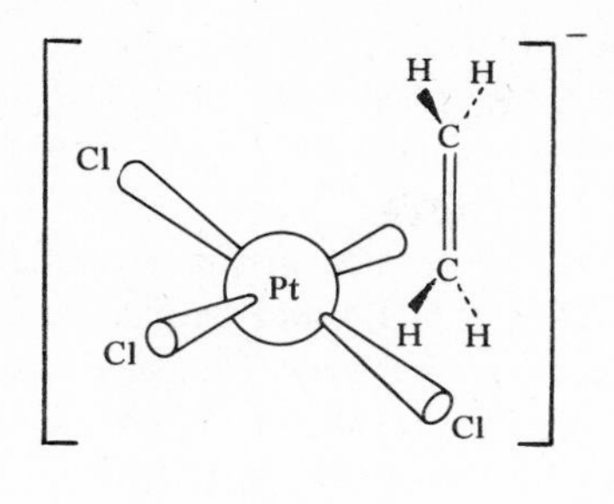

(a)

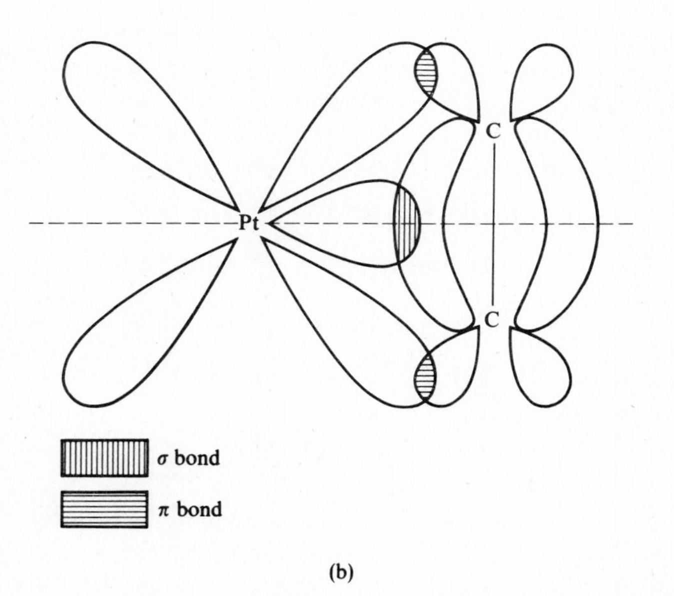

(b)

FIGURE 4-4 Structure *(a)* and bonding *(b)* in the trichloroethyleneplatinate(II) ion.

A wide variety of substances can act as ligands for metal ions, including halide, cyanide, and thiocyanate ions, ammonia, and water. Their common feature, of course, is that they contain an unshared pair of electrons which they can donate to the metal ion. Many very complex ligands are known in which an atom such as nitrogen, oxygen, or sulfur coordinates to the metal ion. Thus proteins can bind metal ions via the nitrogen atoms in the imidazole group of a histidine moiety or via the sulfhydryl group of cysteine, to mention only two possibilities.

Roman numerals are used to specify the oxidation states of metal atoms in complex ions. Thus iron(II) and manganese(IV) designate oxidation states of 2+ and 4+, respectively. The usual rule for assigning the oxidation state of a complexed metal ion is to subtract, algebraically, the sum of the ligand charges from the charge of the total complex. With large ligands, such as proteins, this rule is not convenient, and the oxidation state of the metal ion is usually explicitly stated. By convention, a hydride ligand is assigned an oxidation state of 1−.

As we shall see in the following sections, the reactivity of a ligand may be greatly affected when it is complexed to a metal ion. Conversely, a ligand may have an effect on the reactivity of a metal ion. Consider the series of three half-reactions below.

$$[Fe(CN)_6]^{3-} + e^- \longrightarrow [Fe(CN)_6]^{4-} \qquad E^0 = 0.36 \text{ v}$$

$$[Fe(H_2O)_6]^{3+} + e^- \longrightarrow [Fe(H_2O)_6]^{++} \qquad E^0 = 0.77 \text{ v}$$

$$[Fe(phen)_3]^{3+} + e^- \longrightarrow [Fe(phen)_3]^{++} \qquad E^0 = 1.12 \text{ v}$$

where phen = *o*-phenanthroline = [structure: N N]

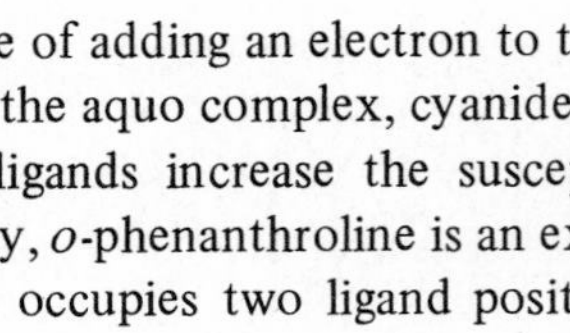

E^0 is a relative measure of the ease of adding an electron to the iron(III) ion in each case; thus, relative to the aquo complex, cyanide ion ligands decrease and *o*-phenanthroline ligands increase the susceptibility of iron(III) to reduction. Incidentally, *o*-phenanthroline is an example of a bidentate ligand, a ligand which occupies two ligand positions of the metal ion via its nitrogen atoms.

4-2 METAL IONS AS SUPERACID CATALYSTS

Introduction

In many metal-ion-catalyzed reactions the role of the metal ion is similar to that of a proton. In Sec. 3-2 we noted that a proton can assist a reaction by drawing electrons to itself, thus weakening the bond to be broken. A positively charged metal ion can perform the same function, but often even more efficiently than the proton because many metal ions can carry more than a single positive charge. Furthermore, in neutral aqueous solution (for example, pH 7) the concentration of a metal ion can easily be 0.1 *M* or more, whereas the hydronium ion concentration is only 10^{-7} *M*. Thus, we can expect that metal ions in neutral solution might be much better catalysts than the hydronium ion. For this reason metal ion catalysts, which operate in the same way as protons, are sometimes called superacid catalysts.

In the remainder of this section we discuss some types of reactions which are susceptible to catalysis by metal ions acting as superacids.

Decarboxylation Reactions

Oxaloacetic acid is one of several acids whose decarboxylation [Eq. (4-1)] is catalyzed by a variety of metal ions. The sequence of steps in this reaction, with Cu^{++} as the catalyst, is shown in Eqs. (4-2) and (4-3).

$$^{-}OOC{-}C({=}O){-}CH_2{-}COO^{-} + H^{+} \longrightarrow {}^{-}OOC{-}C({=}O){-}CH_3 + CO_2 \quad (4\text{-}1)$$

Anion of oxaloacetic acid — Anion of pyruvic acid

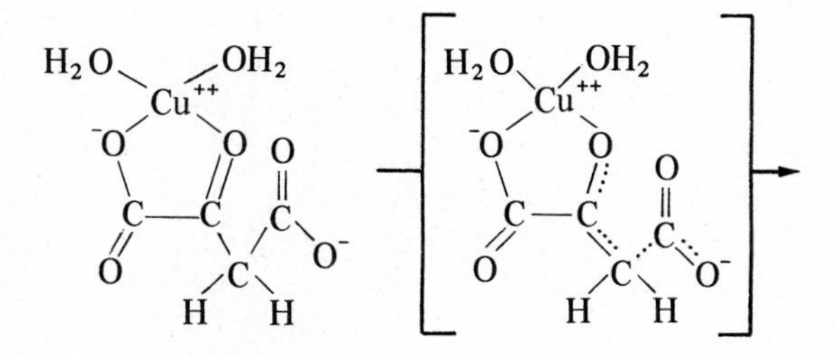

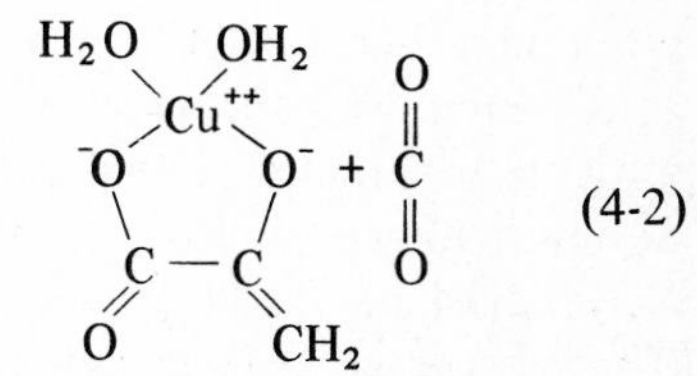

(4-2)

Enol form of pyruvate ion

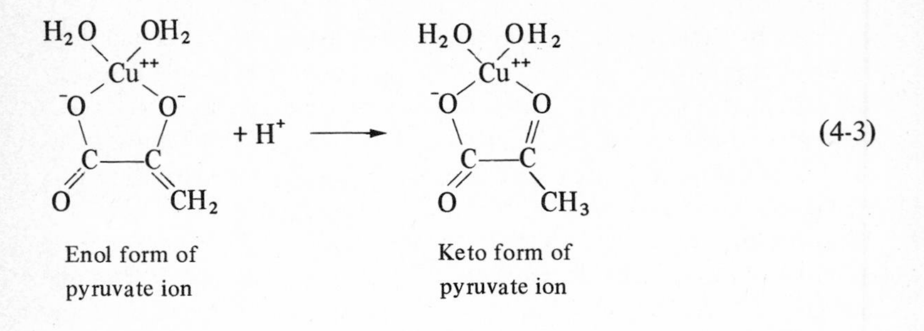

The reactant in Eq. (4-2) is a complex ion formed between the cupric ion and oxaloacetate ion, in which the latter provides two ligands of the complex. Cupric ion generally forms distorted octahedral complexes; the other ligands in this case are, no doubt, water molecules. The reaction proceeds by two steps with the formation of the enol form of pyruvic acid as an intermediate. The term "enol" is used to designate that the molecule contains a carbon-carbon double bond [en(e)] with an alcohol (ol) group on one of those carbons. (If the Cu^{++} were not present, the alcohol group would be an hydroxy group complete with proton.) The second step of the reaction [Eq. (4-3)] is the conversion of the enol form of pyruvic acid to the more stable keto form.

We shall discuss the catalytic effect of the Cu^{++} in this reaction by referring to the hypothetical free-energy diagram of Fig. 4-5. The solid curve is for the uncatalyzed reaction, the dotted for the catalyzed reaction. The uncatalyzed reaction, like the catalyzed reaction, probably also proceeds by two steps: the cleavage of a carbon-carbon bond with the release of CO_2, followed by the conversion of the enol to the keto form of the product. As illustrated in Fig. 4-5, the first of these two steps has the greater free energy of activation and is therefore the rate-determining step. The presence of Cu^{++} as a catalyst changes the mechanism by first forming a complex with the reactant which is more stable than the free oxaloacetate ion; i.e., the complex has a lower free energy, by an amount ΔG_c, than the free ion. If the metal ion had no other effect than to form a complex with the substrate, it would act as an inhibitor for the reaction because it would lower the energy of the initial state without changing the energy of the transition state. In fact, however, the cupric ion is a catalyst for the reaction. This means that $\Delta G_c^{\ddagger}$ must be less than $\Delta G_u^{\ddagger}$. In other words the cupric ion lowers the free energy of the transition state more than it lowers the free energy of the initial state.

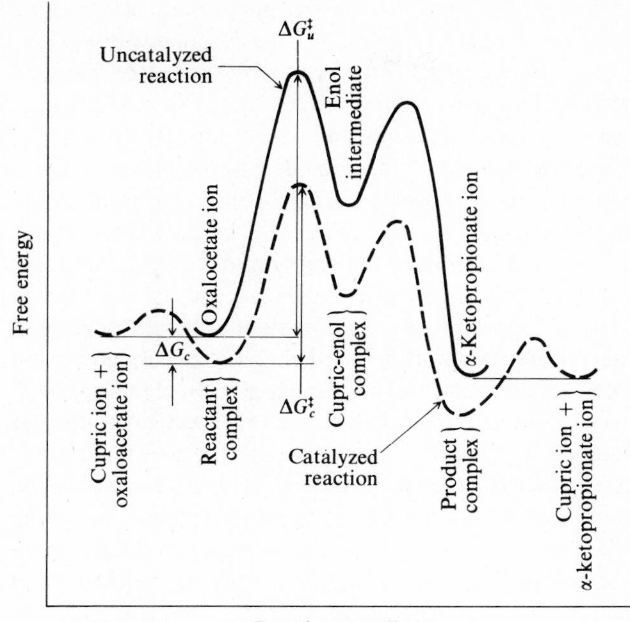

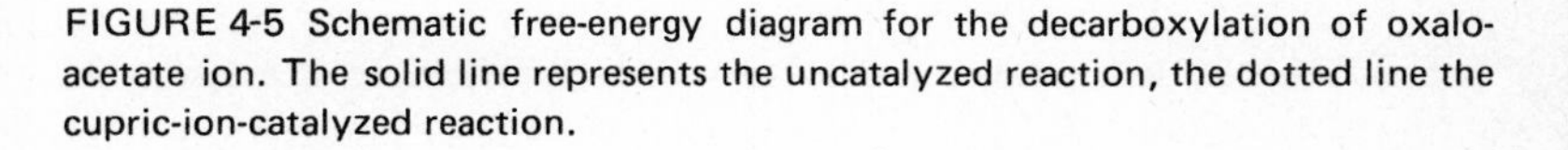
FIGURE 4-5 Schematic free-energy diagram for the decarboxylation of oxalo-acetate ion. The solid line represents the uncatalyzed reaction, the dotted line the cupric-ion-catalyzed reaction.

Why is the cupric ion an effective catalyst? Because of its multiple positive charge, the cupric ion tends to draw electrons toward the keto oxygen atom which is one of its ligands. This is precisely the sort of electronic rearrangement which is required for reaction as shown in the transition state of Eq. (4-2). Thus, it is fundamentally the positive charge on the cupric ion which makes it an effective catalyst. In other words it has the same effect as a proton, but it is more effective because it has a higher charge and can be present at a higher concentration in neutral solution. Thus it is a superacid catalyst.

This discussion suggests several predictions which have been confirmed experimentally. First, other things being equal, the efficiency of a catalyst should increase as its charge increases. Indeed Fe^{3+} is a better catalyst for this reaction than Fe^{++}. Second, if the positive charge of the metal ion is neutralized by the presence of negatively charged ligands [in place of H_2O in Eqs. (4-2) and (4-3)], the metal ion should be less effective as a catalyst. This prediction is also borne out; when acetate ion, which is a ligand for Cu^{++}, is present in the reaction mixture, the rate decreases. Conversely, ligands which enhance the positive charge of the metal ion are observed to increase its effectiveness as a catalyst.

In living organisms there is an enzyme which catalyzes the decarboxylation of oxaloacetate ion. This enzyme utilizes the manganous ion Mn^{++}. The mechanism is probably similar to that of Eqs. (4-2) and (4-3). However, there must be some differences because the enzyme-manganous ion combination causes the reaction of Eq. (4-2) to occur more than 10^8 times faster than manganous ion alone! Obviously, the enzyme plays a very important role. Factors such as those discussed in the previous chapter are no doubt involved. In addition the groups on the enzyme which act as ligands to Mn^{++} [in place of H_2O in Eq. (4-2)] may enhance its positive charge leading to a more effective catalyst, as mentioned before.

Decarboxylation reactions are of general biological importance. Many of the enzymes which catalyze the decarboxylation of amino acids require metal ions to be effective. Metal ions also assist certain coenzymes to bring about decarboxylation reactions, as we shall see in the next chapter.

Amide and Ester Hydrolysis

The hydrolyses of amides and esters are susceptible to the catalytic action of a variety of metal ions. Derivatives of the amino acids are particularly susceptible because the α-amino and the carbonyl oxygen groups are two good potential ligands for complex ion formation.

As an example of this type of reaction we shall consider the complex ion catalyst β-hydroxoaquotriethylenetetraminecobalt(III) which can be abbreviated to $[Co(trien)OH(OH_2)]^{++}$. This is a particularly interesting catalyst because it is selective. In the presence of a polypeptide, it catalyzes the hydrolysis of only the *N*-terminal amino acid from the chain; it does not attack any of the other peptide bonds. Certain enzymes, called aminopeptidases, show this same specificity for *N*-terminal amino acids. Since some of these enzymes also require metal ions, the mechanism of the enzyme catalysis may include aspects of the mechanism we shall now consider.

The mechanism by which $[Co(trien)OH(OH_2)]^{++}$ catalyzes the hydrolysis of *N*-terminal amino acids from polypeptides is illustrated in Eqs. (4-4) and (4-5). For simplicity, alanine is shown as the *N*-terminal residue; R represents the rest of the polypeptide chain.

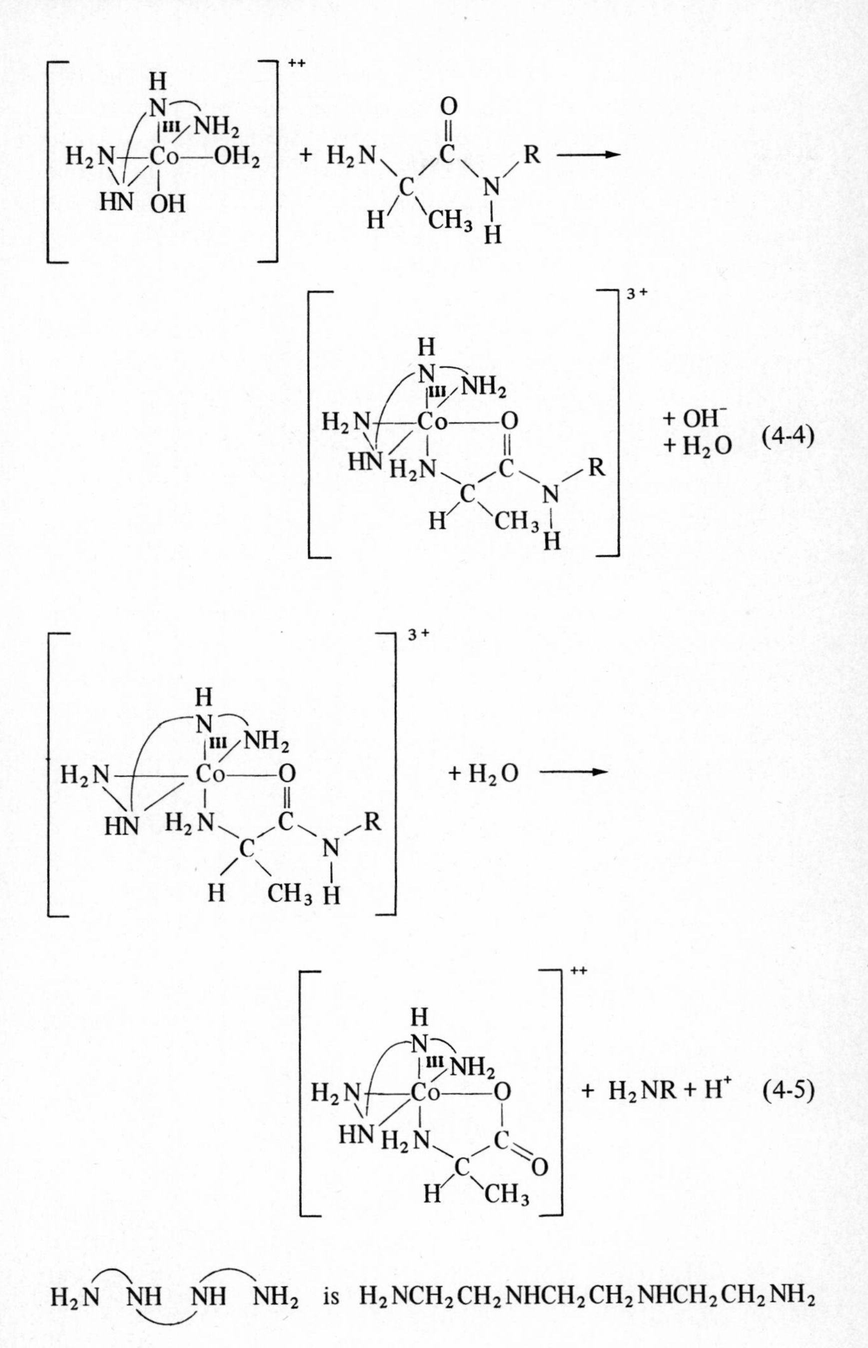
+ OH⁻
+ H₂O (4-4)
+ H₂O
+ H₂NR + H⁺ (4-5)
H₂N NH NH NH₂ is H₂NCH₂CH₂NHCH₂CH₂NHCH₂CH₂NH₂

In Eq. (4-4) the H_2O and OH^- ligands are displaced by the two ligand groups of the *N*-terminal amino acid, alanine. With the carbonyl oxygen atom bonded to the triply positive cobalt ion, the carbonyl carbon loses a large part of its share in the electrons of the double bond. Thus it is much more susceptible to attack by a water molecule than in the uncomplexed state. Again the cobalt ion plays much the same role as a general acid catalyst (Sec. 3-5), but more effectively because of its high charge.

In the example we have just discussed we have spoken of the cobalt complex as a catalyst. Actually it is not a true catalyst since it is not regenerated; more properly it is a promoter of the reaction.

Phosphate Cleavage Reactions

Acetyl phosphate is a mixed anhydride, the two parts of which are derived from acetic acid and phosphoric acid. As an anhydride it hydrolyzes readily in water, though not as fast as acetic anhydride. The mechanism of the hydrolysis reaction differs from that of carboxylic acid anhydrides. The rate-determining step, as shown in Eq. (4-6), is a unimolecular reaction, i.e., only the reactant molecule is directly involved. (We have shown the reaction as it occurs at pH 7 where acetyl phosphate has two negative charges.) The product of Eq. (4-6) is the metaphosphate ion which very rapidly reacts with water to give the orthophosphate ion, Eq. (4-7).

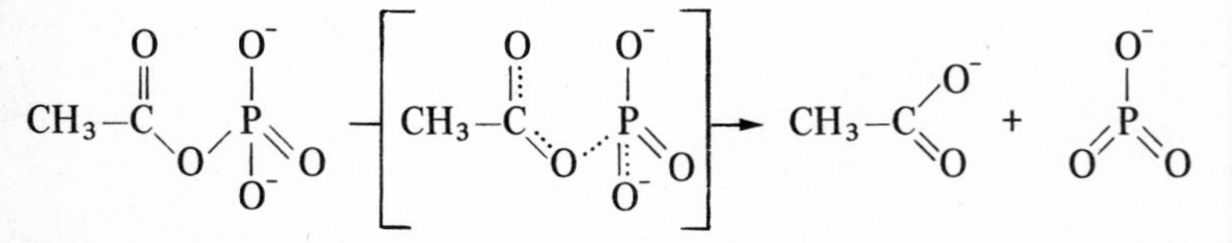

Acetyl phosphate

Meta-
phosphate
ion

(4-6)

$$PO_3^- + H_2O \xrightarrow{\text{fast}} HO\text{–}P(=O)(O^-)_2 + H^+ \qquad (4\text{-}7)$$

Metaphosphate ion

Orthophosphate ion

Both magnesium and calcium ions can catalyze the hydrolysis of acetyl phosphate. The relatively weak complex shown on the left in Eq. (4-8) is believed to form first; its reaction is analogous to Eq. (4-6) but is greatly facilitated by the superacid catalytic effect of the metal ion. The metaphosphate ion reacts again as in Eq. (4-7).

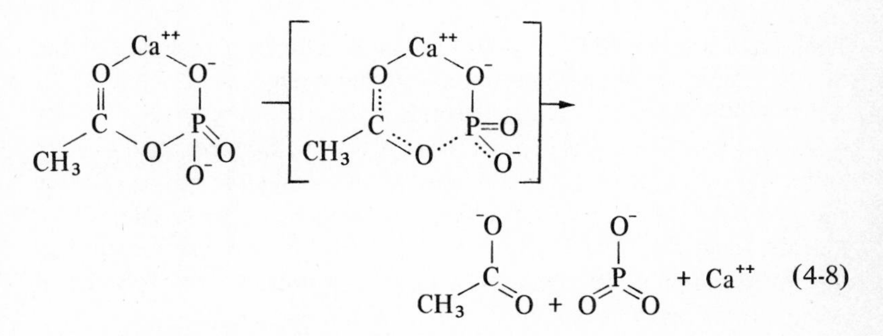

Adenosine triphosphate, usually referred to as ATP, is an exceedingly important metabolite found in living organisms.

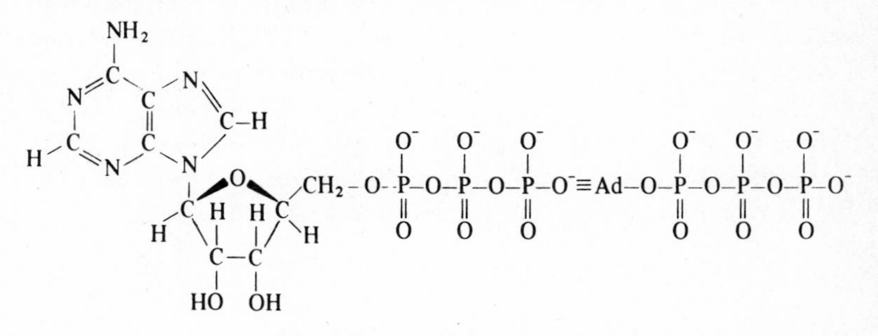

Adenosine triphosphate (ATP)

Almost all of its reactions require the presence of a magnesium ion, which is believed to complex with it and act as a superacid in the same way as in Eq. (4-8). Thus the mechanism for ATP hydrolysis in the presence of Mg^{++} is as shown in Eq. (4-9).

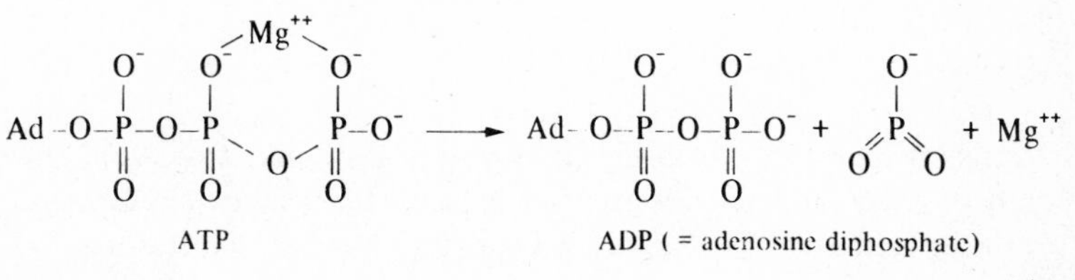

ATP

ADP (= adenosine diphosphate)

(4-9)

The metaphosphate ion can then react with water or any other nucleophile which may be present in the solution.

$$ATP^{4-} + H_2O \longrightarrow ADP^{3-} + HPO_4^{--} + H^+ \qquad (4\text{-}10)$$

$$\Delta G_{pH\ 7,\ 37°C} = -7900 \text{ cal mol}^{-1}$$

The conversion of ATP to ADP plus phosphate ion, as shown in Eq. (4-10), releases a considerable amount of free energy; at pH 7 and 37°C, $\Delta G = -7900$ cal mol^{-1}. (Note that this is the free energy of hydrolysis; it is *not* the free energy of activation, $\Delta G^{\ddagger}$, for the reaction. By definition ΔG equals the free energy of the final state minus the free energy of the initial state. Therefore, the negative value of ΔG tells us that the final state, ADP^{3-} plus HPO_4^{--} plus H^+, has less free energy than the initial state, ATP^{4-} plus H_2O. In other words, a negative value of ΔG indicates that free energy is released in the reaction.)

Applying Eq. (1-15) reveals that the equilibrium constant for the hydrolysis reaction of Eq. (4-10) is 1.6×10^5 *M*, which tells us that the reaction goes essentially to completion. This means that ATP has a high *tendency* to hydrolyze. However, the *rate* of ATP hydrolysis is rather slow even in the presence of Mg^{++} at concentrations found in cells, i.e., the free energy of activation is quite large. Consequently, ATP is sufficiently stable so that it can move from place to place within the living cell and act as a source of chemical free energy. One example of this function is the synthesis of glutamine from glutamic acid.

$$^{-}O{-}\overset{\overset{O}{\|}}{C}{-}\overset{\overset{\overset{+}{N}H_3}{|}}{C}H{-}CH_2{-}CH_2{-}\overset{\overset{O}{\|}}{C}{-}O^- + NH_4^+ \rightleftharpoons$$

Glutamate ion

$$^{-}O{-}\overset{\overset{O}{\|}}{C}{-}\overset{\overset{\overset{+}{N}H_3}{|}}{C}H{-}CH_2{-}CH_2{-}\overset{\overset{O}{\|}}{C}{-}NH_2 + H_2O \qquad (4\text{-}11)$$

Glutamine

$$\Delta G_{pH\ 7,\ 37°C} = 3500 \text{ cal mol}^{-1}$$

The direct reaction shown in Eq. (4-11) cannot produce any significant amount of glutamine because the free-energy change for the overall reaction is unfavorable. The reader can convince himself that this is so by calculating the equilibrium constant for the reaction using Eq. (1-15).

Then it is a simple matter to show that the maximum concentration that can be produced is 3.4×10^{-9} *M*, if 1×10^{-3} *M* is taken as a typical concentration of each of the two reactants; i.e., only 0.00034 percent of the glutamate could be converted to glutamine. However, the enzyme glutamine synthetase can couple (combine) the hydrolysis of ATP with the conversion of glutamate ion to glutamine, as shown in Eq. (4-12), which is the sum of Eqs. (4-10) and (4-11).

$$\text{Glutamate ion} + NH_4^+ + ATP^{4-} \longrightarrow \text{glutamine} + ADP^{3-} + HPO_4^{--} + H^+ \quad (4\text{-}12)$$

$$\begin{aligned} \Delta G &= \Delta G\ [\text{for Eq. (4-10)}] + \Delta G\ [\text{for Eq. (4-11)}] \\ &= (-7900) + (3500) \qquad (4\text{-}13) \\ &= -4400 \text{ cal mol}^{-1} \end{aligned}$$

Since the overall free-energy change for this reaction is large and negative, the equilibrium constant is large and positive. Therefore, the reaction can go essentially to completion.

How does glutamine synthetase manage to couple the two reactions? The reaction is believed to occur in two steps, both of which are catalyzed by the enzyme. First, glutamate ion reacts with ATP, as shown in Eq. (4-14); then the product, glutamyl phosphate, reacts with the ammonium ion, as shown in Eq. (4-15). Both these reactions are catalyzed by the enzyme, which is capable of binding all of the necessary molecules simultaneously. The sum of these two steps is the same as Eq. (4-12); therefore, the free energy calculated in Eq. (4-13) applies to this overall reaction also, and it can go essentially to completion.

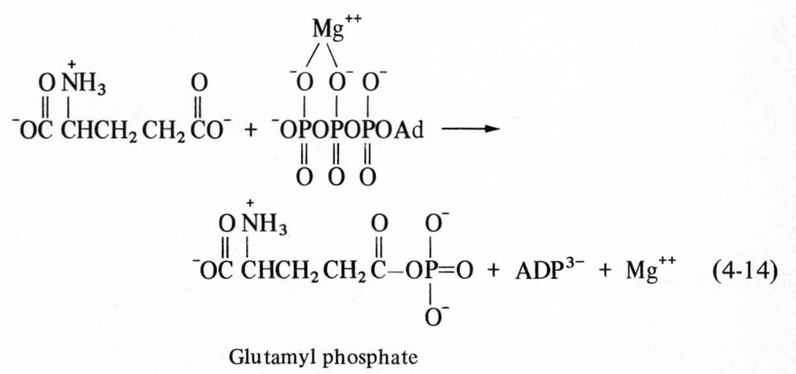

Glutamyl phosphate
(Bound to enzyme)

$$^{-}O\overset{O}{\overset{\|}{C}}\,\overset{\overset{+}{N}H_3}{\overset{|}{C}}HCH_2CH_2\overset{O}{\overset{\|}{C}}-O-\underset{\underset{O^-}{|}}{\overset{\overset{O^-}{|}}{P}}=O + NH_4^{+} \longrightarrow$$

(Bound to enzyme)

$$^{-}O\overset{O}{\overset{\|}{C}}\,\overset{\overset{+}{N}H_3}{\overset{|}{C}}HCH_2CH_2\overset{O}{\overset{\|}{C}}NH_2 + HPO_4^{--} + H^{+} \qquad (4\text{-}15)$$

4-3 METAL IONS AS ELECTRON TRANSFER AGENTS

Redox Reactions

Redox reactions are reactions in which atoms change oxidation state. In

$$Tl^{+} + 2Ce^{4+} \rightleftharpoons Tl^{3+} + 2Ce^{3+} \qquad (4\text{-}16)$$

thallous ion is oxidized by two ceric ions in aqueous solution or, vice versa, two ceric ions are reduced by a thallous ion. By definition, oxidation occurs when the oxidation number (in this case the charge) increases; reduction occurs when the oxidation number decreases. Nontransition metals exhibit, at most, two stable positive oxidation states, e.g., thallium ions can exist in either the 1+ (Tl^{+}) or 3+ (Tl^{3+}) oxidation state. The 2+ oxidation state of thallium is extremely unstable and difficult to form. Consequently, Tl^{+} is a two-equivalent reagent because it can be oxidized only by the loss of two electrons simultaneously. On the other hand, the only stable oxidation states of cerium are 3+ and 4+. Consequently, Ce^{4+} is a one-equivalent reducing agent because it can lose only one electron. Thus the only possible mechanism for Eq. (4-16) is one in which two ceric ions collide simultaneously with one thallous ion. Such a collision is unlikely, as we noted in Chap. 3, particularly so in this case because of repulsions due to charge. Therefore, it is not surprising that the reaction is quite slow.

Manganese, a transition element, can exist in any of the oxidation states from 2+ to 7+, inclusive. Thus it can act as a one-equivalent or a two-equivalent reagent. For this reason, the manganous ion (Mn^{++}) can catalyze the reaction of thallous ion with two ceric ions, by introducing a pathway in which only two-body collisions occur. The mechanism is shown in the following equations.

$$\begin{aligned} Ce^{4+} + Mn^{++} &\longrightarrow Ce^{3+} + Mn^{3+} \\ Ce^{4+} + Mn^{3+} &\longrightarrow Ce^{3+} + Mn^{4+} \\ Tl^{+} + Mn^{4+} &\longrightarrow Tl^{3+} + Mn^{++} \end{aligned} \qquad (4\text{-}17)$$

The sum of these equations gives Eq. (4-16), and Mn^{++} is a true catalyst since it is not consumed.

It is a characteristic property of the transition metals that they are stable in several oxidation states. For this reason, these metals can often act as catalysts in reactions involving electron transfer. Their ability to form complexes with ligands is an important additional property. Coordination can increase the ease of electron transfer. This is illustrated by the catalytic effect of Mn^{3+} on the redox reaction between chlorine and oxalic acid. The most likely mechanism for this reaction is given in Eq. (4-18).

Step 1:

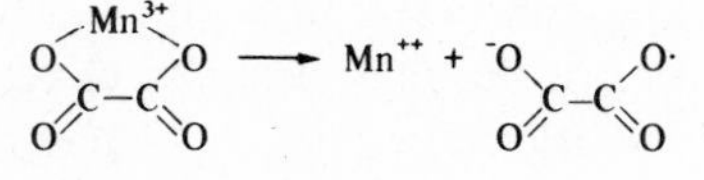

Step 2:

$$Mn^{3+}(C_2O_4) \longrightarrow Mn^{++} + {}^{-}O{-}C(=O){-}C(=O){-}O\cdot$$

Step 3:

$${}^{-}O{-}C(=O){-}C(=O){-}O\cdot \longrightarrow {}^{-}O{-}\dot{C}{=}O + O{=}C{=}O$$

Step 4:

$${}^{-}O{-}\dot{C}{=}O + Cl{-}Cl \longrightarrow O{=}C{=}O + Cl^{-} + Cl\cdot$$

$$(4\text{-}18)$$

Step 5:

$$Cl\cdot + Mn^{++} \longrightarrow Cl^{-} + Mn^{3+}$$

Sum:

$$H_2C_2O_4 + Cl_2 \longrightarrow 2CO_2 + 2Cl^{-} + 2H^{+}$$

Step 1 is the formation of the complex; the other ligand positions of Mn^{3+}, which are not shown, are occupied by water molecules or anions added with the manganese salt. In step 2, which is the rate-determining step for the total reaction, one electron is transferred to Mn^{3+} from one of the negatively charged oxygen atoms leaving that oxygen atom with a free, unpaired electron (indicated by a dot). Such a chemical species is called a *free radical*. It is quite reactive and decomposes rapidly. The free electron, seeking another electron to pair up with, finds its mate in the carbon-carbon bond, as shown in step 3. The dotted line through the middle of that bond indicates that it splits *homolytically*, one electron from the original bond going with each fragment. The new free radical fragment then readily donates its extra electron to a molecule of chlorine as shown in step 4. Finally, the electron transfer of step 5 regenerates the catalyst.

The free radical reactions of steps 3 and 4 of Eq. (4-18) are very fast, i.e., their free energies of activation are small. This is so because an unpaired electron has a natural tendency to pair up with another electron. For this reason the reverse reaction in step 3 would also be fast, except that it is unlikely since it requires the collision of two particles which are at low concentrations. The same reasoning applies to the reverse of step 4.

Because an electron has a small mass, one might naively expect that a simple electron transfer between two metal ions should be very fast. It is known that electrons in atoms or molecules can be excited from a low energy level to a high energy level in less than 10^{-14} s. However, this latter process is quite different from transferring an electron from one ion to another in solution. To discuss the problem concretely let us consider the so-called "exchange" reaction which occurs when a solution containing Fe^{++} ions is added to one containing Fe^{3+} ions. The latter ion was "labeled" by the presence of some of the radioactive isotope of mass number 55. The rate at which radioactivity begins to appear in the ferrous ions is related to the rate of exchange of an electron from Fe^{++} to radioactively labeled Fe^{3+} to give radioactively labeled Fe^{++}.

$$Fe^{++} + {}^{55}Fe^{3+} \rightleftharpoons Fe^{3+} + {}^{55}Fe^{++} \qquad (4\text{-}19)$$

From the experimental results one can calculate that only one out of every ten million collisions results in electron transfer. Clearly, there is a considerable activation energy for this process. Why?

First, in aqueous solution each ion has six water molecule ligands. Thus the two ions cannot get closer together than twice the diameter of a water molecule, about 5 Å (5×10^{-8} cm).

Second, since both ions are positively charged, the repulsion between them will prevent most encounters from being even this intimate.

A third factor is more subtle and very important. The ferric ion, because of its higher positive charge, holds its six polar water ligands more closely to it than does the ferrous ion, as illustrated in a slightly exaggerated schematic way on the left side of Eq. (4-20). Thus, if an electron goes from Fe^{++} to Fe^{3+}, the new species have the wrong ligand-to-metal ion bond lengths. They are too short in the new Fe^{++} complex, too long in the new Fe^{3+} complex. Consequently, the electron will return to its former home. A successful electron transfer will occur only if, at the instant of collision, the Fe^{3+} complex is significantly expanded and the Fe^{++} complex is significantly compressed. These conditions are high energy situations and account for a good part of the high energy of activation.

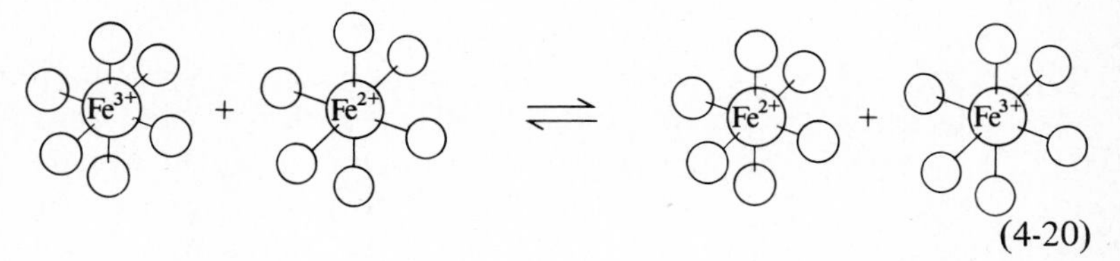

(4-20)

It is this third factor of ligand rearrangement which probably contributes most to the slowness of the exchange reaction between aquated ferrous and ferric ions. In the ferrocyanide [$Fe(CN)_6{}^{4-}$] and ferricyanide [$Fe(CN)_6{}^{3-}$] ions, the difference in ion-ligand distance for the two ions is very small. Thus an electron transfer between these two complex ions should require only little ligand rearrangement. As expected, this exchange reaction is much faster than that between the aquated iron ions.

In summary, although an electron can be transferred between ions very rapidly in principle, the rate of transfer will, in general, be limited by the nature of the ligands about the ion.

Redox reactions are not limited to reactions in which only electrons are transferred from one reagent to another. For example, the disproportionation of hydrogen peroxide is a redox reaction. It may be viewed formally, as shown in Eq. (4-21), but this is not to be taken as the mechanism.

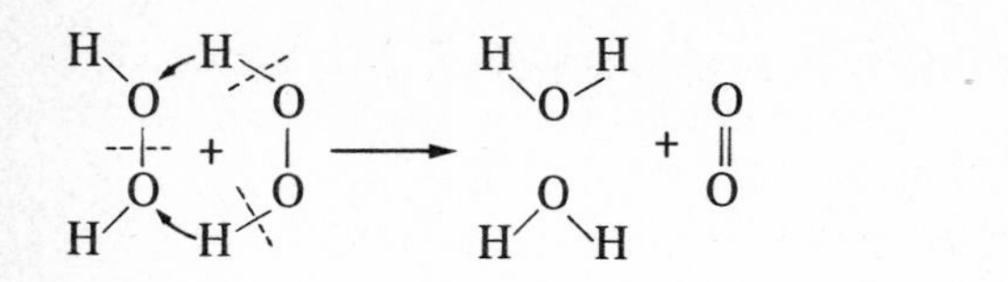

(4-21)

In effect two hydrogen *atoms* are transferred from one molecule to the other. The normal oxidation state of hydrogen is 1+, and so the usual, non-redox, way in which the element hydrogen is transferred from one molecule to another is in the form of protons. Thus a hydrogen atom transfer is a means of transferring one electron from one molecule to another. By the same token the transfer of a hydride ion (H^-) is a two-electron, or two-equivalent, transfer.

Copper as a Redox Catalyst

The cupric ion is a very effective catalyst for one-equivalent redox reactions because it is easily converted to the cuprous ion and back again to the cupric ion by the consecutive gain and loss of one electron. This type of cupric ion catalysis is seen in the reaction of Fe^{3+} and V^{3+} to give Fe^{++} and V^{4+}. The uncatalyzed reaction occurs largely by a simple transfer of one electron from the V^{3+} ion to the Fe^{3+} ion. When cupric salt is added to the solution at a concentration approximately equal to that of the reactants, the rate of the reaction is increased. The mechanism involves the formation of a cuprous ion intermediate, which is the rate-determining step, followed by a fast electron transfer from Cu^+ to Fe^{3+}.

$$V^{3+} + Cu^{++} \longrightarrow V^{4+} + Cu^{+} \qquad (4\text{-}22)$$

$$Cu^{+} + Fe^{3+} \longrightarrow Cu^{++} + Fe^{++} \qquad (4\text{-}23)$$

The cupric ion is a part of the active site in several enzymes, known as oxidases, which catalyze the reaction between O_2 and oxidizable substrates. In some of these enzymes the cupric ion is temporarily reduced to the 1+ oxidation state, analogous to Eq. (4-22). An example is the enzyme which catalyzes the oxidation of hydroquinone by oxygen. This reaction is illustrated in outline in Eqs. (4-24) and (4-25). These are not detailed mechanisms; obviously the first equation includes at least two steps since it involves two cupric ions.

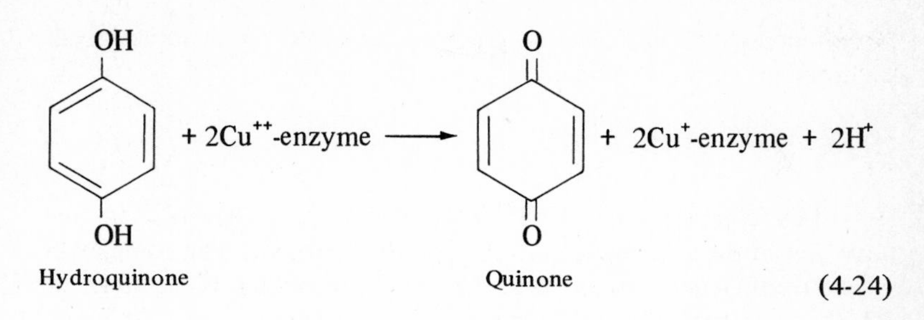

(4-24)

$$4Cu^{+}\text{-enzyme} + O_2 + 4H^{+} \longrightarrow 4Cu^{++}\text{-enzyme} + 2H_2O \qquad (4\text{-}25)$$

Another interesting enzyme is tyrosinase which contains copper in the form of Cu^{+}. Among the reactions it catalyzes is the hydroxylation of the benzene ring in phenylalanine to give tyrosine. It is believed that a molecule of oxygen combines as a ligand to the cuprous ion in this enzyme and that this complex participates in the hydroxylation reaction. The same sort of complex is believed to occur in hemocyanin although no redox reaction occurs. Hemocyanin is an oxygen-carrying protein, not an enzyme. It performs the same function in certain cephalopods, such as crabs and lobsters, that hemoglobin performs in higher animals. As might be expected of a copper complex, oxygenated hemocyanin is blue; these cephalopods are literally the blue bloods of the animal kingdom.

Of the metal-containing enzymes which catalyze redox reactions, the majority utilize either copper or iron at the active site. The latter is probably the most widespread, and we consider it next.

Catalysis by Iron

Oxidation reactions involving molecular oxygen are of vital importance to all higher organisms. Iron is intimately involved in most of the enzymes which catalyze these reactions. For this reason, of the many redox reactions catalyzed by iron (in ionic form), we shall consider examples related to reactions of molecular oxygen only.

The decomposition of hydrogen peroxide in water [Eq. (4-26)] is catalyzed by many transition-metal ions, including ferric ion, which can be added to the solution by dissolving a salt like ferric nitrate. This reaction has been studied for several decades. Its mechanism is still not understood; however, it is generally agreed that free radical intermediates are formed. Of more immediate interest, it appears that ferrous ions are also formed as intermediates. Thus the ferric ion catalysis in this case is reminiscent of the cupric ion catalysis we noted in the previous section.

$$2H_2O_2 \longrightarrow 2H_2O + O_2 \qquad (4\text{-}26)$$

Certain complex ions of Fe^{3+} are much better catalysts than free ferric ion in the decomposition of hydrogen peroxide. The complex in which triethylenetetramine fills four ligand positions is as much as 10,000 times better than the simple ferric ion in this reaction. This case illustrates how a change in ligand can affect the properties of an ion. Not only does the triethylenetetramine ligand enhance the catalytic activity of the ferric ion, it seems to cause the ferric ion to operate by a different mechanism in the decomposition of hydrogen peroxide. Hydrogen peroxide is a weak acid of pK_a 11.7 which ionizes to give the anion ^-OOH. Since ferric ion usually complexes with six ligands, two of these anions can fill the two available ligand positions to give the complex ion shown on the left of Eq. (4-27).

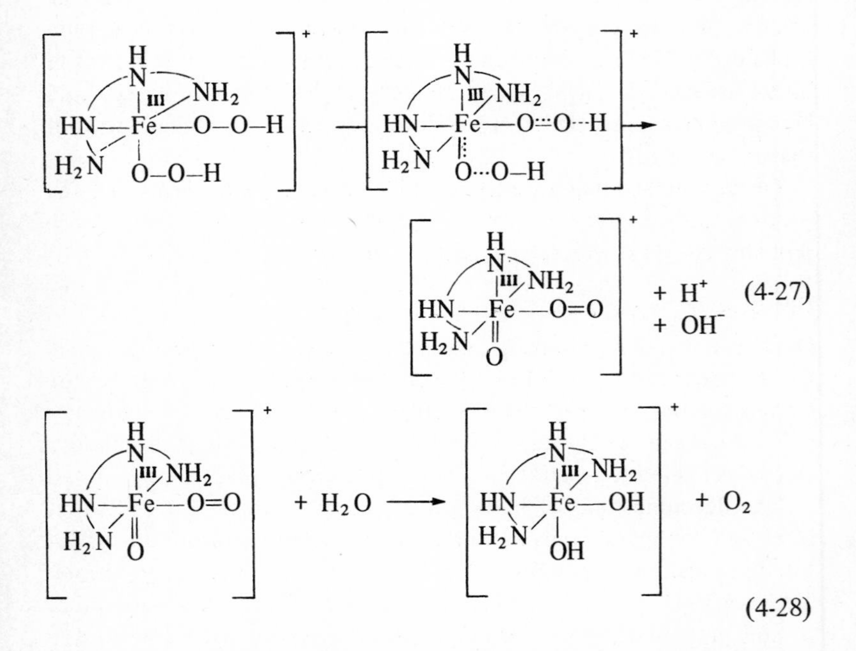

The reaction shown in Eq. (4-27) is assisted by a water molecule, not shown, that accepts the proton liberated. Formally we may assign the following oxidation states in Eq. (4-27): on the left side, Fe has a 3+ oxidation state, and each oxygen is in the 1− oxidation state. On the right side, the O_2 molecule is a neutral ligand, each atom of oxidation state zero; the other oxygen atom has a 2− oxidation state. Consequently, Fe remains in the 3+ oxidation state. In short, the ferric ion mediates the transfer of electrons from one hydroperoxy ligand to the other without undergoing any change itself. Thus the complexed ferric ion plays a dual role: (1) it increases the probability of reaction by bringing the two reacting species together, as ligands; and (2) its electronic makeup is such that it can assist electron transfer between the two ligands. This latter factor is not well understood, but it is real. The triethylenetetramine complexes of a number of other transition metal ions are ineffective as catalysts in the decomposition of H_2O_2. The only exception is the Mn^{++} ion, which, interestingly enough, is isoelectronic with Fe^{3+}, i.e., they each have 23 electrons of which 5 are *d* electrons.

Enzymes which contain iron almost always contain a heme group which acts as a tetradentate ligand for the iron ion. The structure of the heme group varies slightly for different proteins in the nature of the R groups in the structure in Fig. 4-6. One or two of the R groups represent covalent bonds to amino acid side chains in the protein in certain cases. In other enzymes, an amino acid side chain, such as imidazole, holds the heme to the protein by acting as a fifth ligand to the iron ion. An enzyme of this type is catalase, which catalyzes the decomposition of H_2O_2. The heme iron is in the 3+ oxidation state. Its mechanism of action is not clearly understood, but it may be similar to the mechanism we discussed in Eqs. (4-27) and (4-28). One problem, however, is that with the imidazole group from the enzyme acting as a fifth ligand, only one ligand site on the ferric ion is available for reaction with peroxide.

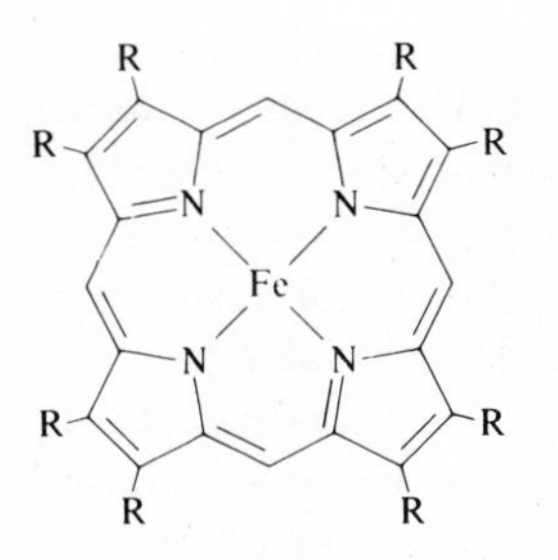

FIGURE 4-6 The iron-heme complex.

Another interesting reaction of iron is the oxidation of ferrous ion to ferric ion by oxygen in water. The mechanism is given in steps 1 and 2 of Eq. (4-29).

Step 1:

$$Fe^{++} + O_2 \longrightarrow \underset{\text{Perferryl ion}}{FeO_2^{\ ++}}$$

Step 2:

$$FeO_2^{\ ++} + H{-}\underset{\substack{|\\H}}{O}{-}Fe^{++} \longrightarrow Fe^{3+}{-}^{-}OOH + HO^{-}{-}Fe^{3+}$$

Step 3:

$$HO^{-}{-}Fe^{3+} + H^{+} \rightleftharpoons Fe^{3+} + H_2O$$

Step 4: (4-29)

$$Fe^{3+}{-}^{-}OOH + H^{+} \rightleftharpoons Fe^{3+} + H_2O_2$$

Step 5:

$$H_2O_2 + 2Fe^{++} + 2H^{+} \longrightarrow 2H_2O + 2Fe^{3+}$$

Sum:

$$4Fe^{++} + O_2 + 4H^{+} \longrightarrow 4Fe^{3+} + 2H_2O$$

(Only the ligands which undergo chemical change are shown.) The perferryl ion formed in step 1 is a ferrous ion with O_2 as one of its ligands. Step 2 shows the transfer of a hydrogen atom, a one-equivalent transfer, from the water ligand of another ferrous ion to the perferryl ion. Simultaneously, each ferrous ion loses one electron to its ligand, giving a hydroperoxy and a hydroxy ligand, respectively. Thus, in step 2 two ferrous ions are oxidized to ferric ions. This is shown more explicitly in steps 3 and 4, which show the dissociation of the two ligands from the ferric ions. The H_2O_2 produced in step 4 is capable of oxidizing two more ferrous ions (step 5). Thus the overall reaction is the oxidation of four ferrous ions by each molecule of oxygen.

The reaction we have just discussed does not involve catalysis. But it is of interest because of its similarity to an important biochemical reaction. All living organisms derive their energy from the oxidation of foodstuffs, such as glucose, by oxygen. The overall breakdown of glucose to CO_2 and H_2O [Eq. (4-30)] occurs via dozens of intermediate reactions, but eventually oxygen is involved.

$$C_6H_{12}O_6 + 6O_2 \longrightarrow 6CO_2 + 6H_2O \qquad \Delta G^\circ = -680 \text{ kcal} \tag{4-30}$$

An example will serve to illustrate the process. The oxidation of malate ion to oxaloacetate ion is one step (from the citric acid cycle) in the breakdown of glucose to CO_2 and H_2O. This overall reaction, shown in Eq. (4-31), releases a considerable amount of free energy.

$$^{-}O{-}\overset{O}{\overset{\|}{C}}{-}CH_2{-}\underset{H}{\underset{|}{\overset{OH}{\overset{|}{C}}}}{-}\overset{O}{\overset{\|}{C}}{-}O^{-} + \tfrac{1}{2}O_2 \longrightarrow {}^{-}O{-}\overset{O}{\overset{\|}{C}}{-}CH_2{-}\overset{O}{\overset{\|}{C}}{-}\overset{O}{\overset{\|}{C}}{-}O^{-} + H_2O \tag{4-31}$$

By carrying out the reaction in several steps, the living cell is able to harness the reaction to produce three molecules of ATP from ADP and phosphate ions, thereby storing a major part of the available free energy for future use in the cell.

In the first step of this sequence [Eq. (4-32)], the enzyme malate dehydrogenase catalyzes the transfer of a hydride ion (H^-) from the malate ion to the coenzyme nicotinamide adenine dinucleotide (NAD^+).

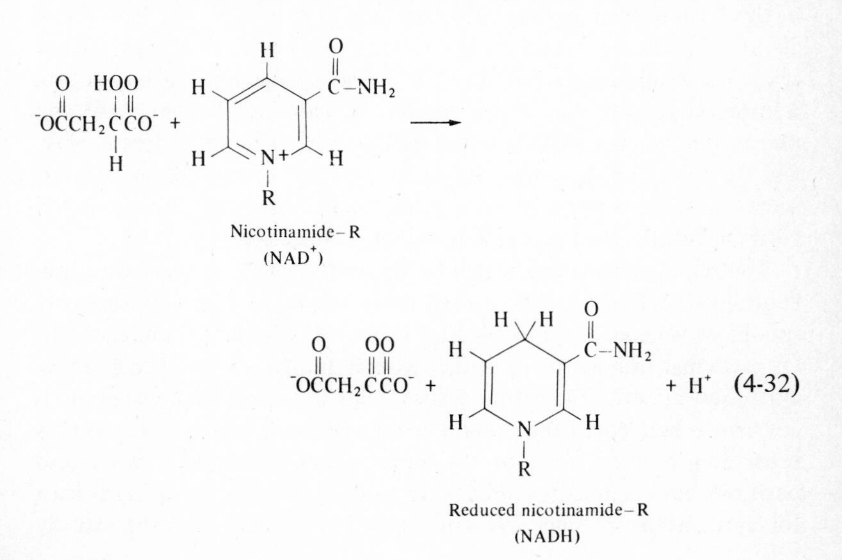

Nicotinamide–R (NAD^+)

Reduced nicotinamide–R (NADH)

Thus two electrons have been transferred from the malate ion and have reduced NAD^+ to NADH. Now follows a series of enzyme-catalyzed electron-transfer reactions from one electron carrier to another, involving about eight carriers. This series of reactions is often called the *electron transport chain*. In each step the electron acceptor is more readily reduced than the electron acceptor of the preceding step; i.e., each reaction has a negative $\Delta G^{\circ\prime}$, the change in free energy at pH 7 in water. In three of these steps that have rather large decreases in $\Delta G^{\circ\prime}$, the enzyme catalyst can function only if both ADP and phosphate ion are present. At the same time that the enzyme catalyzes the electron-transfer reaction, it also brings about the combination of ADP and phosphate ion to give ATP. The first such reaction occurs in the second step of the sequence, the reoxidation of NADH back to NAD^+ with flavin as the electron acceptor. (Flavin is another coenzyme which is tightly bound to the enzyme catalyzing the reaction.) It is as if the enzyme says to the two electrons on NADH: "You want to go to flavin where you would be at a lower energy, but despite your rather high energy, you don't have enough energy to get over the mountain (free energy of activation barrier) to join up with flavin. Only I can lead you through that treacherous mountain pass which does not require so much initial energy. But I can do so only if you will help out my ADP friend here. The energy you will lose is more than enough to help him unite with this phosphate ion. Since you would release the energy in the form of heat anyway, you two may as well help each other. How about it?" The electrons, having neither free will nor feelings, cannot reject the proposition. Since the free energy available from the electron transfer is greater than that required to form ATP from ADP and phosphate ion, the total reaction results in a decrease in free energy, and so it proceeds spontaneously. The presence of the enzyme catalyst causes this reaction to proceed more rapidly than the direct reaction between NADH and flavin that does not produce ATP.

The series of reactions which occurs in the electron transport chain is outlined in Fig. 4-7. (We should point out that this is a considerably simplified version of a process that is not yet completely understood.) Of particular interest to us in this section are the several cytochromes, cyt b, cyt c_1, etc. These are hemoproteins in which the heme group is covalently bonded to the protein via two of the R groups in Fig. 4-6. A ferric ion is complexed to the heme group. The other two ligand positions are occupied by imidazole residues from the protein, at least for cytochrome c. Since cytochromes b, c_1, c, and a are successively

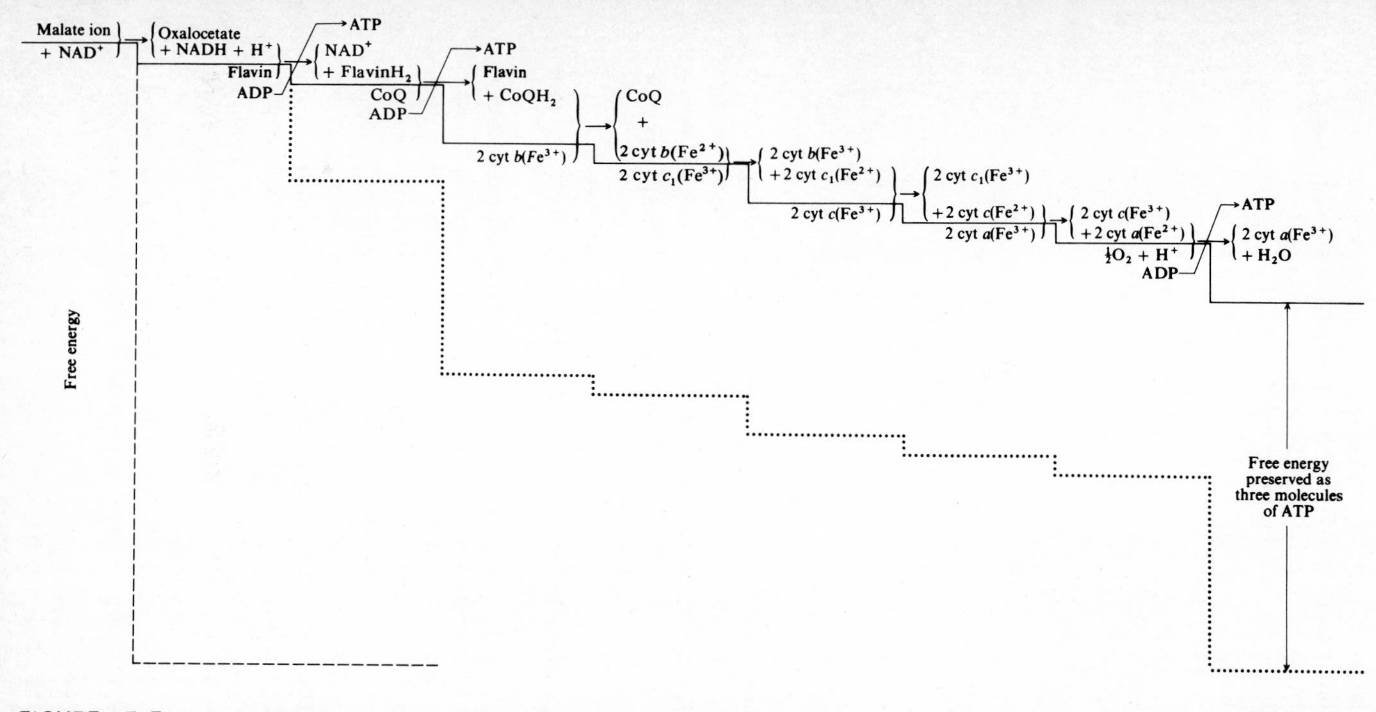

FIGURE 4-7 Free-energy changes in the oxidation of malate ion to oxaloacetate ion by molecular oxygen. The actual pathway in the electron-transport system of the cell is shown as a solid line; CoQ is coenzyme Q; cyt b, cyt c_1, etc., are cytochromes b, c_1, etc. The dotted line shows the free-energy changes in the various steps if no ATP were formed; approximately half the total free energy released when malate is oxidized to oxaloacetate is preserved in the form of three ATP molecules. The dashed line shows the change in free energy when malate is oxidized to oxaloacetate directly in one step, as in Eq. (4-31).

better electron acceptors, there must be slight differences in the way the ligands about the ferric ion affect its reactivity.

Finally, we should observe two things about Fig. 4-7. First, the net result of all the reactions is the same as in Eq. (4-31). All the intermediate compounds end up unchanged, having gone through the cycle, oxidized form → reduced form → oxidized form. Thus all these intermediates are catalysts. Second, the free energy changes illustrate the storage of free energy by the synthesis of ATP. Each time a molecule of ATP is produced, the free energy decrease for the step is less than it would be if no ATP were formed. Since the enzymes for these steps can operate only when the starting materials for ATP are present, the alternative pathway shown as a dotted line in Fig. 4-7, in which ATP is not formed, cannot take place.

4-4 METAL ION CATALYSIS OF POLYMERIZATION REACTIONS

Types of Polymerization Reactions

A polymer is a large molecule made by joining together a large number of small molecules or monomers. For example, the plastic polystyrene consists of large molecules prepared from the monomer styrene.

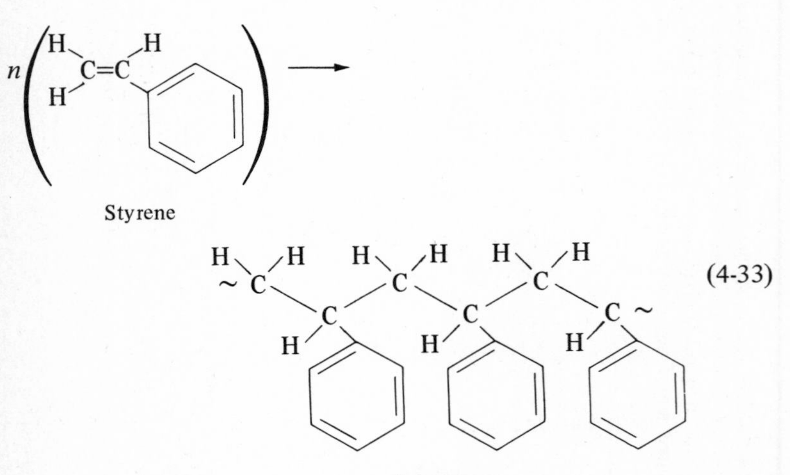

Polymerization reactions are of two types—condensation and addition. A typical addition polymerization is the styrene reaction mentioned above. In such reactions the monomer is the repeating unit in the polymer. The vast majority of addition polymerization reactions utilize unsaturated monomers, generally vinyl compounds. (Substituted ethylene compounds, such as styrene, are called *vinyl compounds.*) Through the conversion of a double bond into a single bond, these unsaturated monomers can become completely incorporated into polymers. In contrast, condensation polymerizations involve the elimination of a small molecule, such as water or hydrochloric acid, in the reaction between monomers. An example is the preparation of nylon.

$$n\mathrm{H_2N(CH_2)_6NH_2} + n\mathrm{HO{-}\overset{\overset{\displaystyle O}{\|}}{C}(CH_2)_4\overset{\overset{\displaystyle O}{\|}}{C}{-}OH} \xrightarrow{280^\circ\mathrm{C}}$$

$$\left[\sim \mathrm{\overset{\overset{\displaystyle H}{|}}{N}(CH_2)_6\overset{\overset{\displaystyle H}{|}}{N}{-}\overset{\overset{\displaystyle O}{\|}}{C}(CH_2)_4\overset{\overset{\displaystyle O}{\|}}{C}} \sim\right]_n + 2n\mathrm{H_2O} \qquad (4\text{-}34)$$

We shall consider addition reactions exclusively.

Addition polymerizations take place by the stepwise addition of monomer units to a growing polymer molecule which is in a reactive form. Polymer growth can occur either by free radical or by ionic polymerization. We shall wait until later to discuss how these reactive species are prepared.

A hydrocarbon free radical is a molecule in which one carbon atom has a single unpaired electron instead of a fourth covalent bond. This is a very reactive species which tends to *add* to another molecule as in the following addition polymerization reaction.

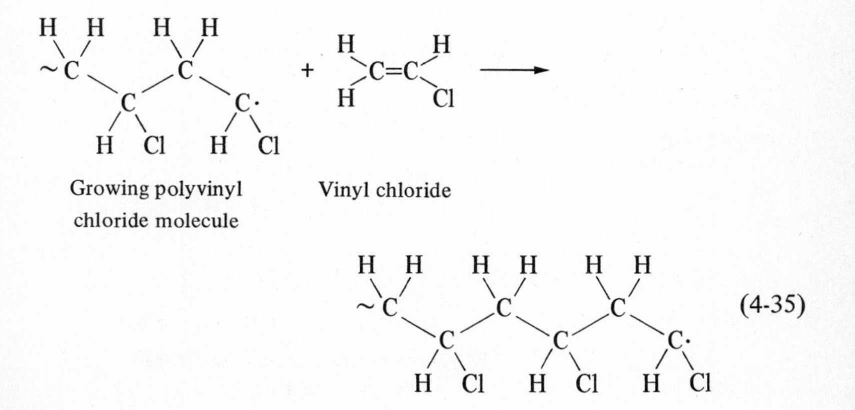

Growing polyvinyl chloride molecule

Vinyl chloride

(4-35)

The unpaired electron pairs up with one electron from the double bond of the monomer leaving one unpaired electron at the new end of the polymer molecule. Thus the process can be repeated continuously until the free radical is destroyed in some way, or the monomer is totally consumed.

In ionic polymerizations the polymer grows in a manner directly analogous to free radical polymerization. The growing polymer molecule has either a positive (carbonium ion) or a negative (carbanion) charge on the carbon at the growing end of the molecule. The charge is preserved after each monomer addition, as illustrated in the following examples.

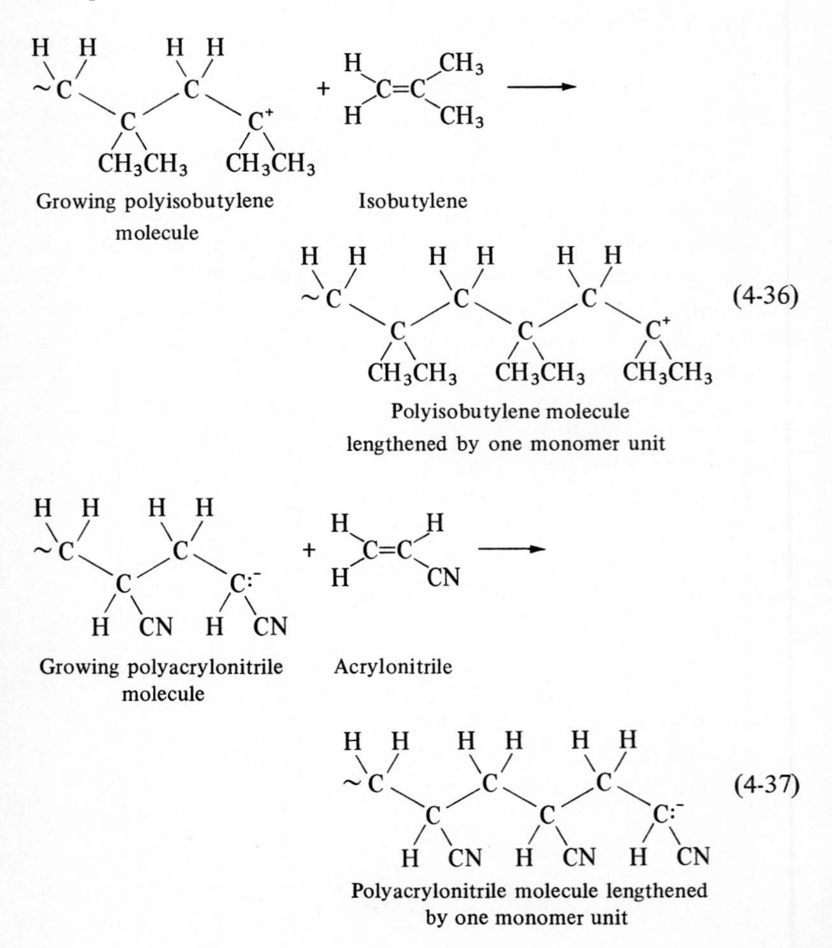

The mechanism of polymerization, whether free radical, carbonium ion, or carbanion, depends on the nature of the substituent in the vinyl monomer. As we have noted several times before, a good intermediate in any chemical reaction must balance high ease of formation with high reactivity. On the one hand, if an intermediate is very reactive it is usually difficult to form. On the other hand, an intermediate which is easily prepared is often not very reactive. In either case, the overall reaction via that intermediate is slow. Normally, carbonium ions and carbanions, such as those derived from ethane (Fig. 4-8), are difficult to form and highly reactive. Thus for ionic polymerization reactions to occur, the intermediate ionic species must be stabilized so that their formation is facilitated. Consider the growing polyisobutylene carbonium ion [Eq. (4-36)]. It has two methyl groups bonded to the positively charged carbon atom. A methyl group is more electron-releasing (less electronegative) than a hydrogen atom. Therefore, the positively charged carbon atom in polyisobutylene has a greater "share" in its bonding electrons than does the corresponding atom in the carbonium ion of ethane. In other words the positive charge in the polyisobutylene ion, being partly shared by the adjacent methyl groups, is more diffuse (more *delocalized*) than that in the ethyl carbonium ion. Consequently, the former ion, although still quite reactive, is more stable than the latter. In fact, it is sufficiently stable so that it achieves a proper balance between high ease of formation and high reactivity. Even at -100°C, isobutylene will completely polymerize to high molecular weight polyisobutylene in a few seconds when the reaction is properly initiated.

From the converse of the preceding discussion, it is clear that an electron-withdrawing group, such as the cyano group [Eq. (4-37)], will stabilize a carbanion. Vinyl monomers which do not form relatively stable ionic species will generally polymerize by the free radical mechanism. Some substituents, such as the phenyl group, are capable of stabilizing both carbonium ions and carbanions. Thus styrene [Eq.

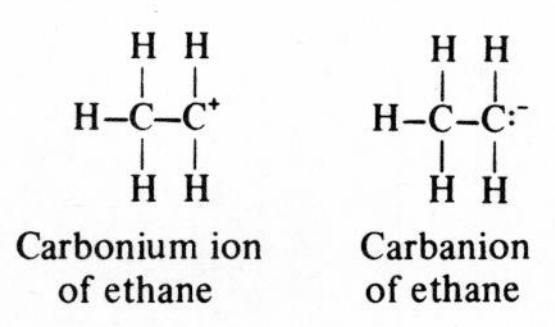

FIGURE 4-8

(4-33)] can polymerize by either of the ionic mechanisms as well as by the free radical mechanism. The actual mechanism of the reaction then will depend upon how the reaction is initiated, as we shall now see.

Chain Reactions

Addition polymerizations are chain reactions. Once a growing free radical or ionic species has been started, it will propagate itself by reacting successively with monomer molecules until the reaction is terminated in some way. Thus there are three stages in a chain reaction: initiation, propagation, and termination. We have just discussed the propagation step in vinyl polymerization, and we turn now to the initiating and terminating processes.

A very common initiator for free radical addition polymerizations is benzoyl peroxide. It can be made to decompose into two benzoyloxy free radicals either by heating it or adding certain catalysts. The free radicals produced [Eq. (4-38)] can start polymer chains by reacting with monomer molecules [Eq. (4-39)] and the chain reaction proceeds to the propagation stage. The reaction is terminated if the free-radical chain center is destroyed.

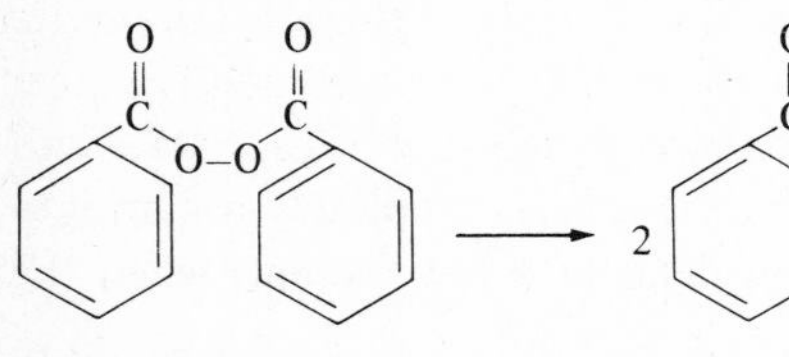

(4-38)

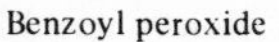

Benzoyl peroxide Benzoyloxy radical

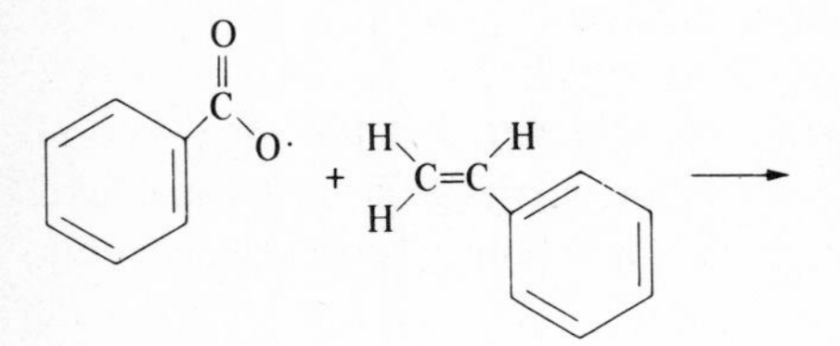

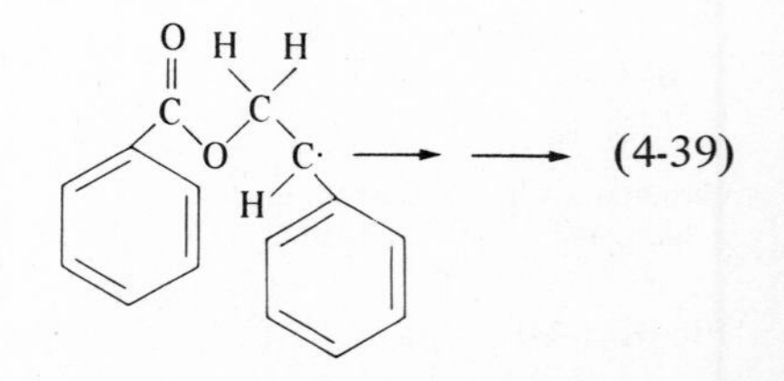

(4-39)

If two free radicals come together and form a covalent bond between themselves by sharing their unpaired electrons with each other, both chain reactions are stopped and the length of the completed polymer molecule is the sum of their two lengths.

Alternatively, inhibitors can react with the growing polymer radical and stop the reaction. Oxygen (O_2), which contains two unpaired electrons and is in a sense a stable free radical, can play such a role, as can quinone and a number of other compounds. Such inhibitors may sometimes be added to control the length of the polymer molecules; however, they are usually undesirable. Inhibitors have been very useful in determining the mechanisms of addition polymerization reactions. For example, if quinone addition causes a decrease in reaction rate, the reaction is of the free radical type, whereas no effect implies that the reaction does not involve free radicals. Since the number of growing polymer chains is usually very small, the addition of a very small amount of inhibitor can have a pronounced effect on a free radical polymerization reaction.

The principles of initiation and termination of ionic polymerization reactions are the same as for free radical reactions. The isobutylene polymerization [Eq. (4-36)] can be initiated by the addition of a Lewis acid such as titanium tetrachloride which reacts with a solvent molecule.

$$\underset{}{TiCl_4} + \underset{\text{(e.g., ethylene)}}{RH} \longrightarrow H^+TiCl_4R^- \tag{4-40}$$

The strong acid thus formed protonates an isobutylene molecule [Eq. (4-41)], and the reaction is started.

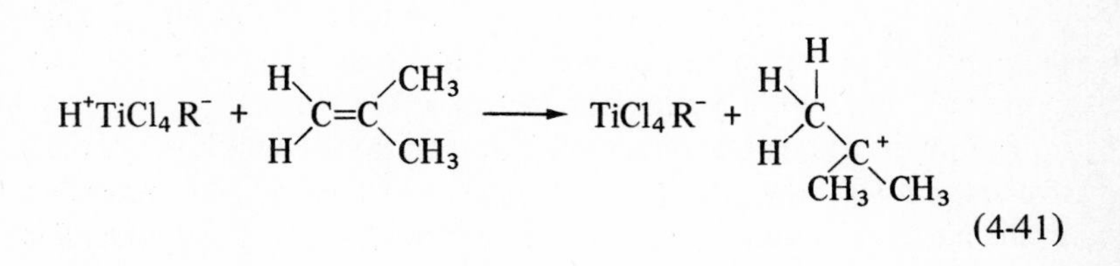

(4-41)

The chain can be terminated by the presence of any species that reacts with carbonium ions, including water among other things. Obviously, the solvents used in this reaction must be both anhydrous and nonbasic. Two other modes of termination are shown in Eq. (4-42), one of which is essentially the reverse reaction of Eqs. (4-40) and (4-41).

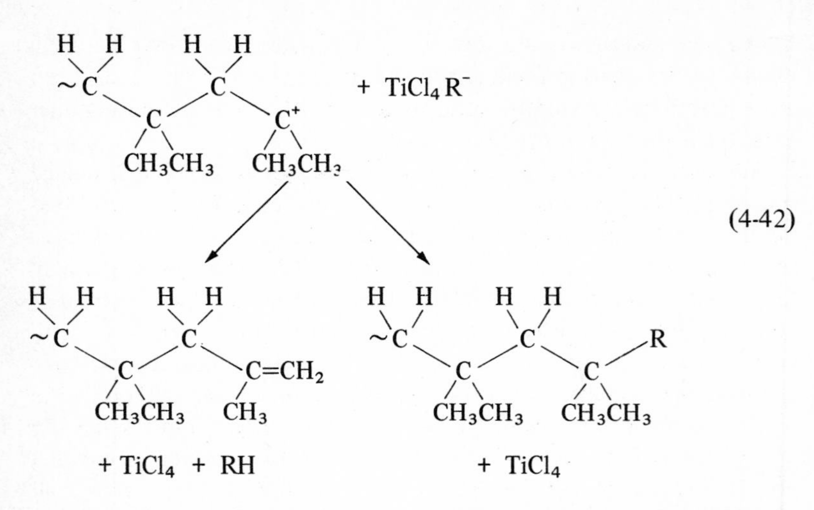

Carbanion chain reactions can be initiated by a strong base. For example, acrylonitrile polymerization is initiated in liquid ammonia by adding an amide salt such as sodamide.

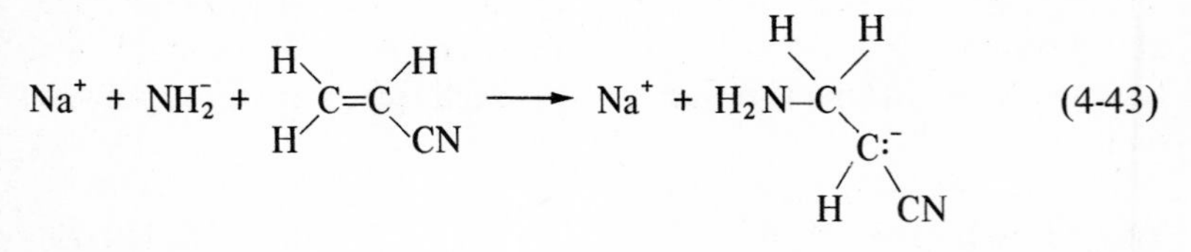

Termination most commonly occurs by a transfer of a proton to the carbanion from a solvent molecule (NH_3) to regenerate the amide ion (NH_2^-).

Sometimes growing polymer molecules are not terminated, but growth stops simply because all the monomer has been consumed. Such polymers are called *living polymers*; addition of more, or a different, monomer will result in further polymer growth.

The properties of a polymeric material (many of which are plastics) depend, among other things, on the length of the polymer molecules. Even under carefully controlled conditions, there is considerable variation in polymer length. The average length will depend upon: (1) the amount of initiator used–the less used, the fewer the number of chains started and the longer each can grow before the monomer is consumed; and (2) the rate of the termination reaction–if the number of chain terminations per second is low, the molecules on the average will be longer than when the termination rate is faster.

Stereochemistry of Polymer Molecules

In Chap. 2 we noted that optical isomers are possible when a compound contains an asymmetric carbon atom, i.e., a carbon atom with four different groups bonded to it. (Optical isomers have the property of rotating the plane of plane polarized light.) A section of polypropylene [a polymer prepared from propylene, Eq. (4-44)] is shown in Fig. 4-9*b*.

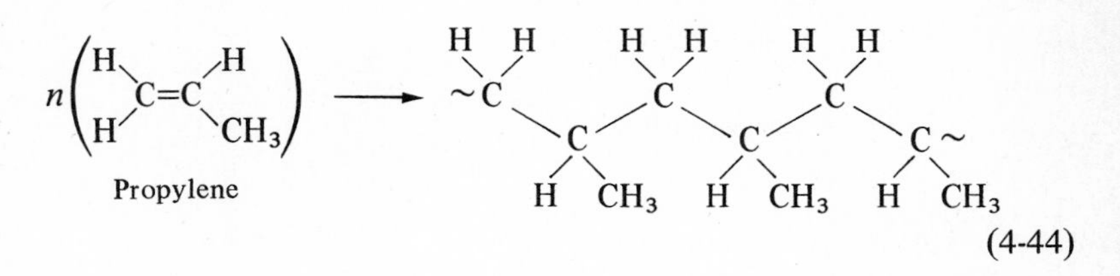

Every second carbon atom in the backbone of this molecule (starred in Fig. 4-9*b*) is asymmetric. However, since two of the substituents, being long alkyl chains differing only in length, are very similar, the polypropylene isomer shown in Fig. 4-9*b* has little effect on polarized light.

A much more important consequence of the asymmetry is the effect it has on the crystallizability of polypropylene. Linear polyethylene, which has no alkyl substituents along the chain (Fig. 4-9*a*), can form crystals in which the long molecules lie compactly side by side somewhat like uncooked spaghetti. If the methyl substituents in polypropylene are randomly oriented in all directions, the extended molecules cannot lie compactly and therefore cannot form crystals. However, if the methyl substituents are oriented in a *regular* way, the molecule is called *stereoregular* and it can lie compactly alongside other such molecules to give crystalline material. Figure 4-9*b* and *c* shows the two possible types of such regularity: (1) an *isotactic* polymer, in which all substituents point in the same direction; (2) a *syndiotactic* polymer, in which the groups alternate in the direction they point. If the orientation is random, the molecule is said to be *atactic* (Fig. 4-9*d*).

For the purposes of illustration, the molecules are drawn in extended zigzag form with the backbone carbon atoms all lying in the plane of the paper. This is the normal conformation of the molecules in crystalline polyethylene. However, in isotactic polypropylene such a conformation is impossible because the methyl groups are too bulky. (The ball-and-stick representations used in Fig. 4-9 do not accurately reflect the actual size of the atoms and methyl groups.) Consequently, the molecule takes up a spiral conformation (Fig. 4-9*e*) which allows

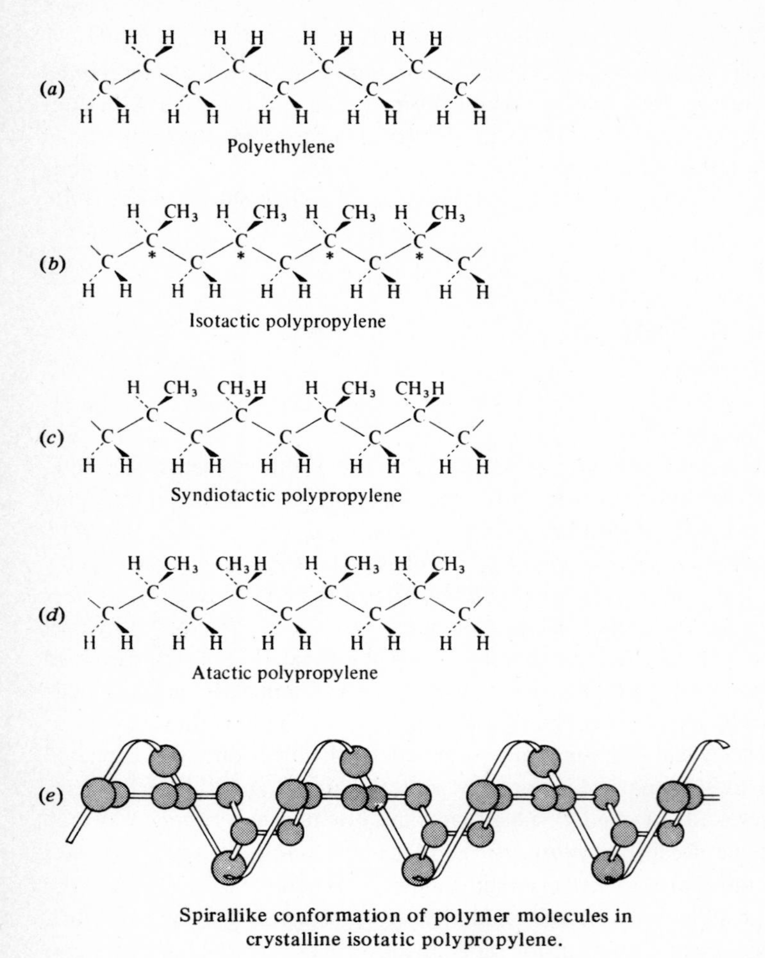

Spirallike conformation of polymer molecules in crystalline isotatic polypropylene.

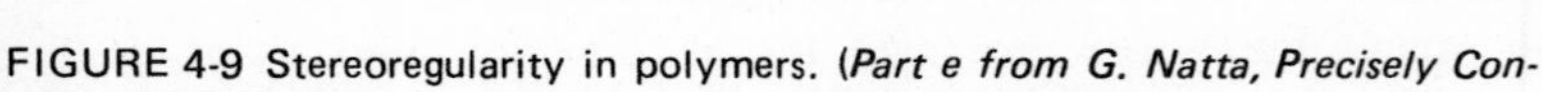

FIGURE 4-9 Stereoregularity in polymers. (*Part e from G. Natta, Precisely Constructed Polymers, Scientific American, p. 39, August, 1961.*

the molecules in a crystal to lie compactly with some intermeshing, somewhat like headless, threaded bolts lying side by side in a box.

Polymerizations Catalyzed by Alkyllithium Compounds

Butyllithium (Fig. 4-10) is the most commonly used alkyllithium catalyst in polymerization reactions. The carbon-lithium bond in these

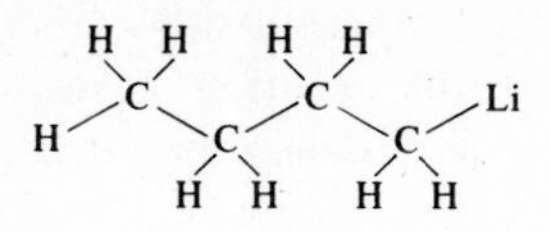

FIGURE 4-10 Butyllithium.

compounds is a highly polar covalent bond, the lithium atom carrying a partial positive charge, the carbon atom a partial negative charge. This bond is highly reactive; therefore alkyllithium compounds combine readily with substances, such as water, oxygen, carbon dioxide, acids, etc., and must be protected from them.

In addition polymerizations catalyzed by butyllithium, the lithium atom always remains bonded to the growing end of the chain. Thus each monomer unit becomes *inserted* between the carbon-lithium bond, as shown in Eq. (4-45) for the initiation step in the polymerization of styrene.

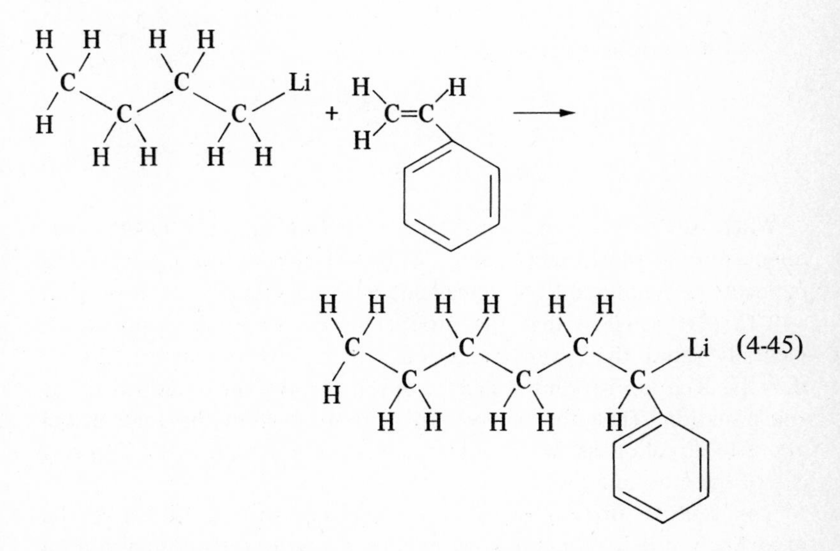

Subsequent propagation steps occur in the same way. Equation (4-46) shows a rather detailed description of the mechanism for this reaction. First the partially positive lithium atom is attracted to the π-electron cloud of the styrene double bond and a vacant lithium orbital overlaps with the electrons (step 1). The positive charge on the lithium atom is highly concentrated because lithium is so small; consequently, the styrene double bond is readily distorted, and the reaction passes through the transition state (step 2) to give a growing polymer lengthened by one monomer unit.

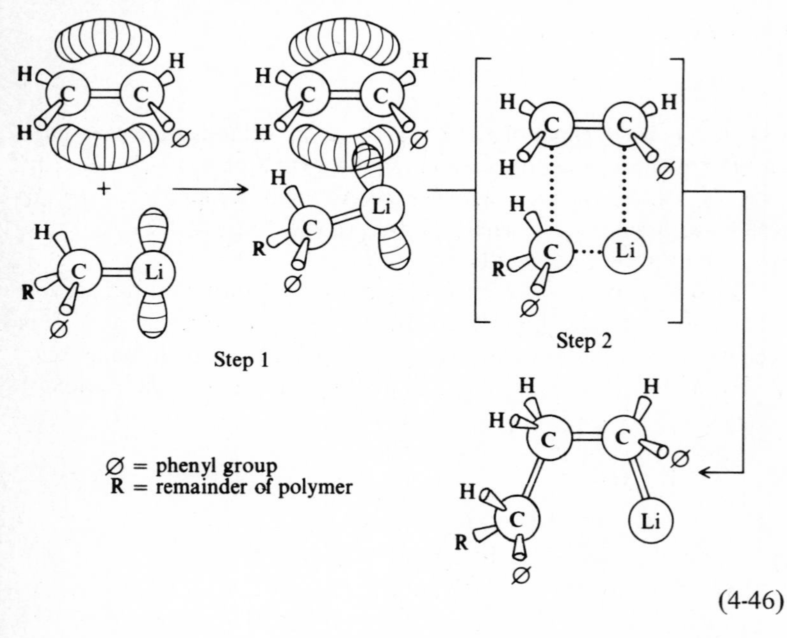

(4-46)

When all propagation steps take place with precisely the same orientation of substituents, as in Eq. (4-46), the resulting polystyrene molecule is syndiotactic in agreement with experiments performed at -40°C. (To visualize that the product in Eq. (4-46) is syndiotactic, mentally rotate the lower left carbon with its substituents by 120° so that the R group is pointed slightly upward to give the extended zigzag conformation. Then the phenyl group points behind the page, in the opposite direction as the phenyl group on the carbon at the growing end of the molecule.)

The relative orientation of the molecules in Eq. (4-46) is the most likely one for the following reasons. Consider the terminal carbon on the growing chain. Of its three substituents (not counting the lithium atom), two, the R and phenyl groups, are very bulky. Thus the CH_2 group of the styrene monomer approaches this carbon atom from the side of the small hydrogen substituent [from the top in Eq. (4-46)]. Two possible orientations of styrene remain, the one with the phenyl group pointing out of the page (as shown), or the one with the phenyl group pointing behind the page. The former orientation involves some interference between the monomer phenyl group and the phenyl group on the terminal carbon of the growing polymer; the latter orientation involves interference between the monomer phenyl and polymer R groups. If the latter orientation were the more favorable energetically, isotactic polymer would result. However, since the R group is bulkier than a phenyl group, the former orientation is of lower energy and syndiotactic polymer is produced at −40°C. Organic chemists would say that the orientation shown in Eq. (4-46) has the least *steric hindrance*, or is sterically most favorable. However, the energy difference between the two orientations of the styrene molecule is small, and as the temperature is raised, collisions become sufficiently energetic so that a considerable fraction of the reactions involve the higher energy orientation. Thus the polystyrene product prepared at 0°C is atactic.

We should add here that other factors may have a bearing on the mechanism of butyllithium-catalyzed reactions. There is some evidence that two or more growing polymer chains form some sort of complex, and this may affect the structure of the product. Also, the nature of the solvent affects the reaction. If an ether is used as solvent instead of a hydrocarbon, the polystyrene product at −40°C is atactic under otherwise identical conditions. Our understanding of these reactions is still not complete.

In the preceding discussion we have referred to butyllithium as a catalyst. Strictly speaking, it is not a catalyst but a promoter, since one molecule is consumed for every molecule of polymer made. However, if an average polymer contains, say, 2,000 monomer units per molecule, only 0.0005 mole of butyllithium is consumed per mole of monomer. Because such small amounts bring about reactions which otherwise would not readily occur, it is customary to refer to these substances as catalysts.

A large number of unsaturated compounds undergo addition polymerization in the presence of alkyllithium catalysts. A particularly interesting case is the polymerization of isoprene [Eq. (4-47)]. Several different polyisoprenes are possible. Focusing on carbon atoms 1 and 2, we may view isoprene as a vinyl compound with methyl and ethylene substituents on carbon-2. Thus 1,2 polymerization will give rise to polymers which can be isotactic, syndiotactic, or atactic. Similarly, 3,4 polymerization can occur.

Another polyisoprene is one in which carbon atoms 1 through 4 become part of the polymer backbone, as shown here.

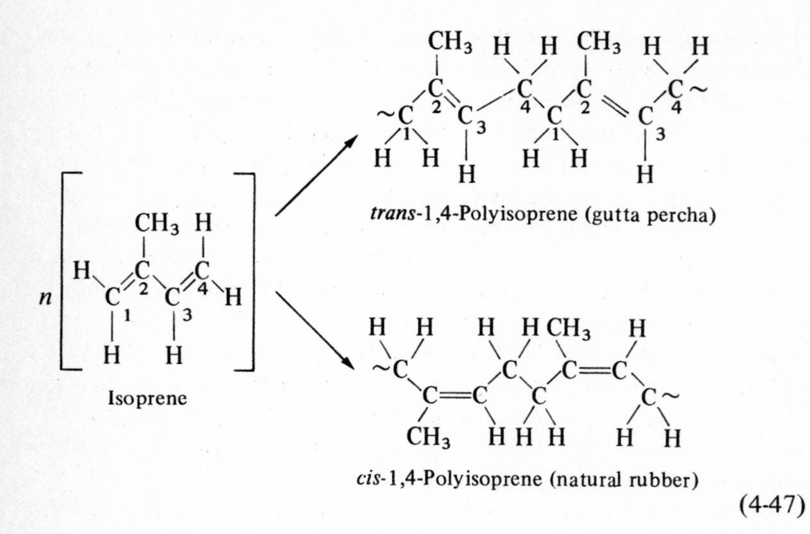

(4-47)

Note that one monomer double bond is lost during polymerization, while the remaining double bond has changed location. The presence of a double bond in the polymer gives rise to geometrical (cis-trans) isomers. If the polyisoprene backbone has the trans configuration about all the double bonds, it is the naturally occurring compound gutta percha, which when vulcanized forms a hard, tough rubber. If the polyisoprene has the cis configuration at all the double bonds, it is identical with natural rubber. In 1956 scientists at the Firestone Tire and Rubber Co. in Akron, Ohio reported the synthesis of *cis*-1,4-polyisoprene using an alkyllithium catalyst. One year earlier scientists at Goodyear announced that they had also successfully made synthetic "natural" rubber using a quite different solid catalyst system now known as a Ziegler-Natta catalyst.

Polymerizations Using Ziegler-Natta Catalysts

In 1963 the Nobel Prize in chemistry was awarded jointly to Karl Ziegler of Germany and Giulio Natta of Italy. Their discoveries and studies of polymerization catalysts between 1952 and 1955 laid the basis for a quantum leap forward in the preparation of commercially important high polymers. Ziegler discovered that when triethyl aluminum and titanium tetrachloride were poured together in a hydrocarbon solvent (such as diesel oil), they formed a brown insoluble solid which catalyzed the polymerization of ethylene to polyethylene.

This polymerization reaction took place readily at 70°C and 1 atm of ethylene. In contrast, the commercial process in use at that time required a temperature of about 200°C and a pressure of 1,000 atm or more. The advantage of the milder conditions with Ziegler's catalyst is largely balanced economically by the greater cost of preparing it. However, it soon became apparent that Ziegler's catalyst produces linear polyethylene molecules which crystallize to give bulk polyethylene of superior strength. In contrast, other polyethylenes were highly branched, therefore generally less crystalline and less strong.

The story of the discovery of Ziegler's catalyst exemplifies the fundamental scientific principles of careful observation and scientific detective work. One day a student of Ziegler's observed that a reaction of ethylene with the catalyst triethylaluminum, which usually gave a mixture of hydrocarbon products, this time gave only pure 1-butene. In tracking down the cause for this change in catalyst behavior, the following probable sequence of events was uncovered. The autoclave had most recently been used for a different reaction using a metallic nickel catalyst. Subsequent cleaning with nitric acid oxidized the nickel to Ni^{++} which was incompletely rinsed out. A final washing with a phosphate-softened water solution formed insoluble nickel phosphate which adhered to the walls of the autoclave. Further studies with compounds of nickel and other transition metals intentionally added showed that they did indeed bestow upon triethylaluminum the ability to dimerize ethylene. But the biggest surprise was yet to come. Experiments with zirconium and titanium compounds added to triethylaluminum gave astoundingly good yields of polyethylene. This is the sort of discovery which keeps a chemist working day and night testing other compounds to find the best combination. As Ziegler says, "In the course of a few dramatic days, yes, even hours, [these experiments] led to the discovery of the new polyethylene process."

When Ziegler's discovery became known in the scientific community, polymer chemists all over the world, as well as Ziegler's research group, began experimenting with this catalyst, changing conditions of preparation, using different transition metal compounds in place of $TiCl_4$, and different Group III metal alkyls in place of triethylaluminum, etc. Indeed, many other combinations are catalytically active. In general, titanium compounds may be replaced by compounds of other early members in the transition series, such as zirconium and vanadium. Triethylaluminum may be replaced by other alkyl compounds of many of the elements in Groups I to III. Solvents are usually hydrocarbons, but in any case they must be free of water.

In 1954 Natta, using monomers other than ethylene, discovered that a Ziegler-type catalyst effected the polymerization of propylene [Eq. (4-44)], giving a product which was crystalline. This was astounding because a crystal could form only if all the long polypropylene molecules were lying compactly together, which in turn would occur only if all the methyl substituents were spatially oriented in a regular way. Natta's x-ray analysis of these crystals showed that the polypropylene was indeed isotactic. Thus catalysts were available (now known as Ziegler-Natta catalysts) which could introduce stereoregularity into a polymer molecule. Of course enzymes have been doing this for a long time, but these were the first man-made catalysts having a wide range capability of this kind.

A knowledge of the mechanism of action of Ziegler-Natta catalysts is of particular interest because, as always, such knowledge will facilitate the search for new, perhaps more efficient, catalysts. One of the systems which has proved most amenable to the investigation of its mechanism is the polymerization of propylene using a mixed titanium trichloride-triethylaluminum catalyst. In heptane solvent, the triethylaluminum dissolves; the titanium trichloride does not, but remains as a crystalline solid on whose surface are "active sites" where polymerization occurs. In crystalline titanium trichloride each titanium atom has six chlorine atom near-neighbors arranged in an octahedral array; each chlorine atom has two near-neighbor titanium atoms. However, on the surface some titanium atoms have only five chloride ligands; these constitute the potential active sites.

The dissolved triethylaluminum plays the role of initiator by alkylating the active site titanium ion and removing one chlorine ligand as shown in Eq. (4-48).

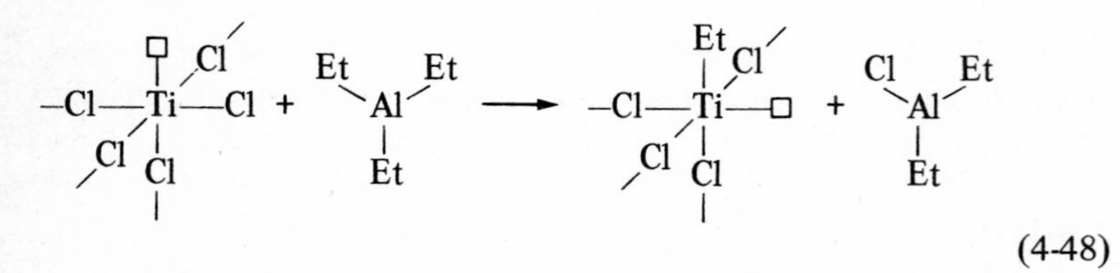

(4-48)

where Et = ethyl group

□ = ligand vacancy

The alkylated titanium surface complex produced in this way lies slightly below the surface with the vacant ligand position near the surface (Fig. 4-11*a*). Propagation of the alkyl chain now proceeds. A molecule of propylene forms a π complex to titanium in the vacant ligand position as shown in Fig. 4-11*b*, the original alkyl ligand reacts with the complexed monomer [first step of Eq. (4-49)], and the now extended alkyl group moves back to the original alkyl ligand position [second step in Eq. (4-49)].

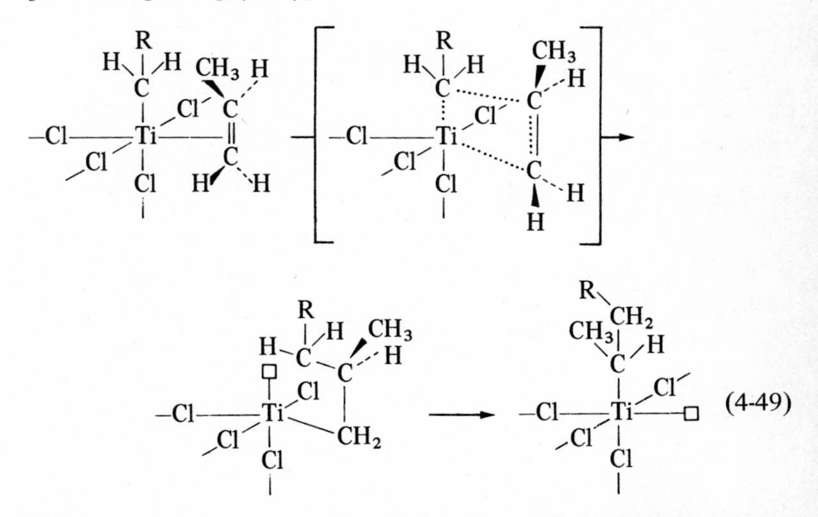

This reaction between two adjacent ligands is called a *ligand migration reaction*.

The reaction is stereospecific because each new propylene monomer (1) always reacts from the same ligand position, and (2) can fit into that ligand position in only one orientation owing to the presence of other surface chloride ions surrounding the hole. The propagation step presumably can occur only if the monomer double bond is parallel with the titanium-carbon bond as in Fig. 4-11*b*. Although the monomer can complex to the alternate ligand position [second-to-last structure in Eq. (4-49)], the surrounding surface atoms hinder it from assuming this parallel orientation. If monomer does complex to this position, polymer growth at this site stops until the monomer dissociates and the alkyl group can move back to its proper position.

Several possible termination reactions have been suggested, one of which is the release of the growing polymer by formation of a titanium hydride [Eq. (4-50)] which can then start a new polymer.

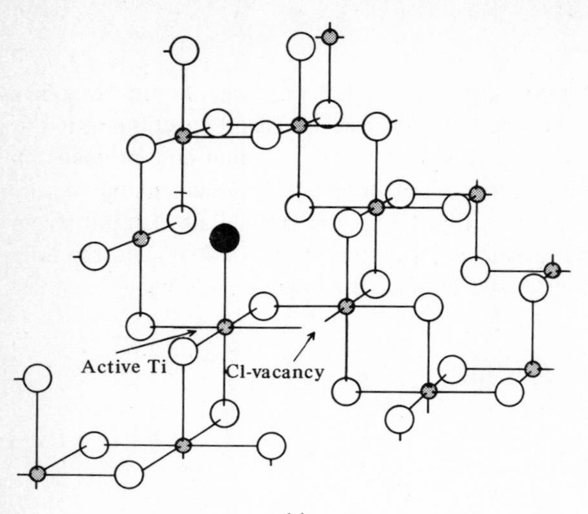

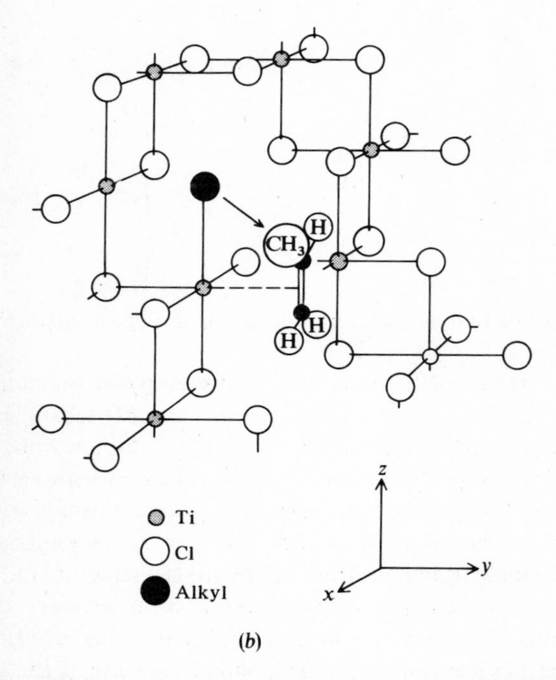

FIGURE 4-11 Surface lattice structure of titanium trichloride showing (*a*) active site titanium ion with a ligand vacancy, and (*b*) a propylene molecule complexed to the titanium ion of the active site. [*From P. Cossee, Stereoregularity in Heterogeneous Ziegler-Natta Catalysis, Transactions of the Faraday Society, 58, 1226 (1962)*.]

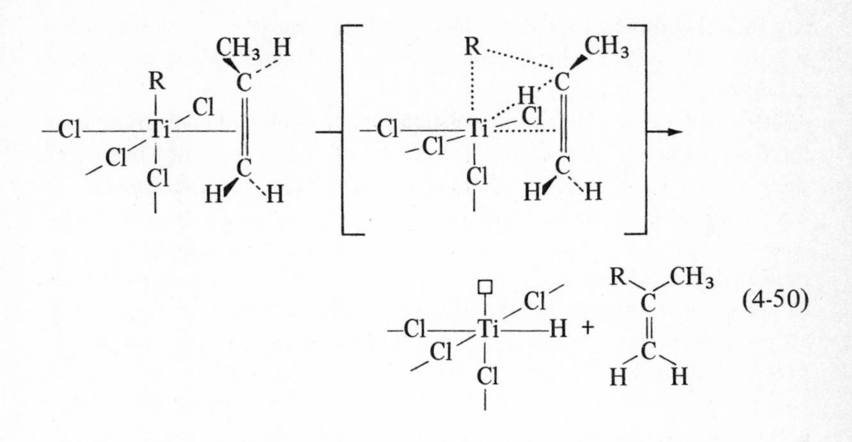

The titanium trichloride acts as a true catalyst in this reaction since it is not consumed; some triethylaluminum is used up in the initiation process, and during the course of the polymerization more of it may regenerate "poisoned" active sites by reacting with and removing the poisoning ligand.

What are the features of Ziegler-Natta catalysts that make them so effective? First, triethylaluminum is a very reactive alkylating agent, consequently it readily brings about the initiation step of Eq. (4-48). Second, titanium (and other near-neighbor transition metals) happens to have the correct electronic structure to form a strong covalent bond to an alkyl group which, however, is made labile when an alkene becomes a neighboring ligand, so that the ligand migration reaction in Eq. (4-49) is extremely facile.

Finally, although attention has been centered on the polymerization of ethylene and propylene, many other monomers can be polymerized by Ziegler-Natta catalysts. For example, Natta has prepared four polybutadienes: isotactic and syndiotactic 1,2-polybutadiene, *trans*-1,4-polybutadiene, and *cis*-1,4-polybutadiene. The last differs from natural rubber [*cis*-1,4-polyisoprene, Eq. (4-47)] only in lacking a methyl group, and in fact makes an excellent synthetic rubber.

4-5 OTHER HOMOGENEOUS METAL ION CATALYST SYSTEMS

Since about 1950, fundamental research in inorganic chemistry has become very popular. Out of this intensified interest have come many new metal-complex catalysts which catalyze reactions in homogeneous solution. The rest of this chapter will deal with some examples of these catalysts.

In many cases heterogeneous catalysts for the same reaction have long been known, but their mechanisms of action have not been well understood. In general it is easier to determine the mechanisms of action of homogeneous than of heterogeneous catalysts for several reasons: (1) A homogeneous metal ion catalyst is usually a small molecule containing only one metal ion, whereas a heterogeneous catalyst, even though finely powdered, contains thousands of metal atoms per particle; (2) thus, a homogeneous metal ion catalyst can be prepared pure, i.e., it contains only one type of metal ion and each catalyst molecule will have the same catalytic activity, whereas a heterogeneous catalyst has surface atoms which may differ in catalytic activity due to differences in oxidation state, closeness to surface, nature of surrounding atoms, etc. In the following we shall emphasize homogeneous metal ion catalysis.

The "Oxo" Synthesis

Complexes between metals and carbon monoxide are called metal carbonyls. In the "oxo" reaction a cobalt carbonyl complex catalyzes the reaction between an alkene, carbon monoxide, and hydrogen [Eq. (4-51)].

$$\text{R–CH=CH}_2 + \text{H}_2 + \text{CO} \longrightarrow \text{R–CH}_2\text{CH}_2\overset{\overset{\displaystyle\text{O}}{\|}}{\text{C}}\text{H} \qquad (4\text{-}51)$$

This is a reaction of considerable industrial importance and one of the few such which utilize a homogeneous catalyst. Under the common reaction conditions of 120°C and 100 to 300 atm, the alkene and/or the product aldehyde are liquids and dissolve the catalyst.

The active form of the catalyst is cobalt hydrotricarbonyl, which is present at low concentrations in equilibrium with cobalt hydrotetracarbonyl. Since cobalt is less electronegative than hydrogen, the assigned oxidation states are 1− for H, 1+ for Co, and 0 for the carbon monoxide ligands. The reaction is quite complex, but the following mechanism, based on considerable study of this and related reactions, seems plausible. This mechanism has some similarities to previous reactions.

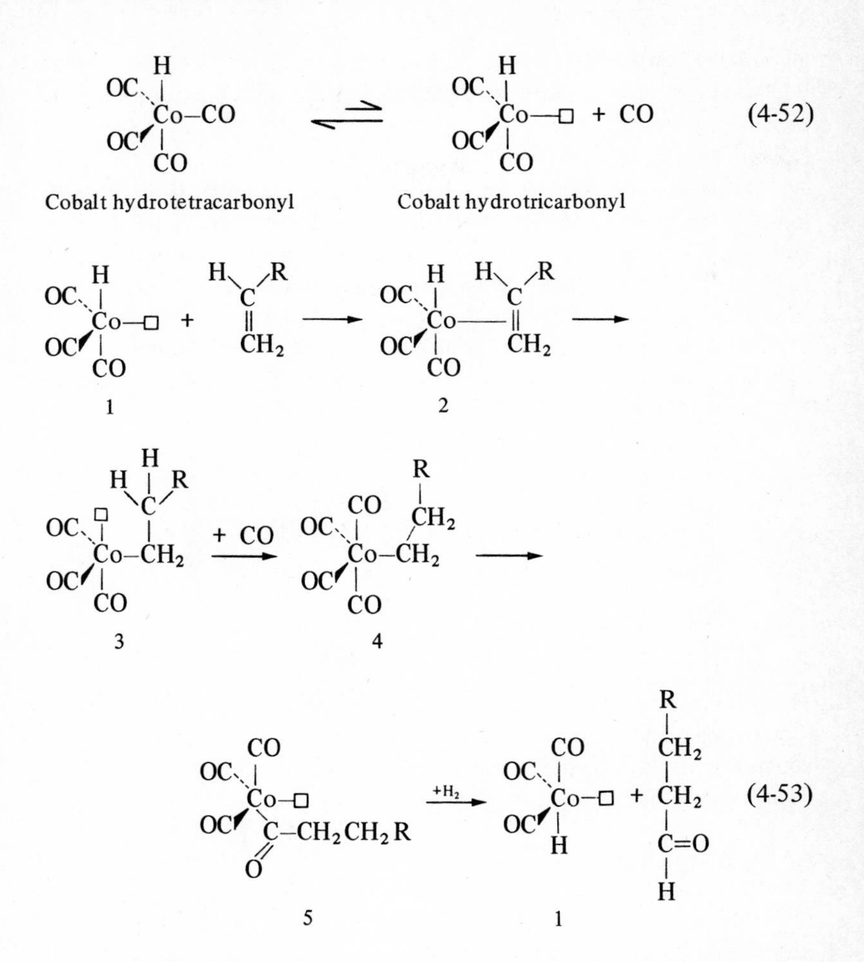

A π complex is formed (species 2) between the alkene and cobalt followed by a migration of the hydrogen atom from cobalt to the alkyl chain to give an alkylated cobalt (species 3). After another molecule of CO combines to give species 4, a second ligand migration of the alkyl group to a carbon monoxide ligand gives species 5, which reacts with hydrogen to regenerate the catalyst and release the aldehyde product. As in the Ziegler-Natta catalysts, an important function of the metal ion is to bring the two reactants into a cis (adjacent) arrangement.

With similar catalysts and slightly altered conditions, other reactions can be carried out. For example, a nickel carbonyl catalyst in acidic aqueous solution under a high pressure of carbon monoxide converts alkenes to carboxylic acids. The oxo synthesis itself, at somewhat higher temperatures, gives alcohols rather than aldehydes as products.

Hydrogenation of Alkenes

A number of stable metal ion complexes containing hydride ion as a ligand have been prepared in recent years. Other less stable hydride complexes are believed to occur as intermediates in the catalytic hydrogenation of alkenes by metal ion complexes, such as tris-(triphenylphosphine)chlororhodium(I) (Wilkinson's catalyst) and ruthenium(II) hexachloride. The latter compound catalyzes the hydrogenation of fumaric acid to succinic acid [Eq. (4-54)], a reaction which has been well studied.

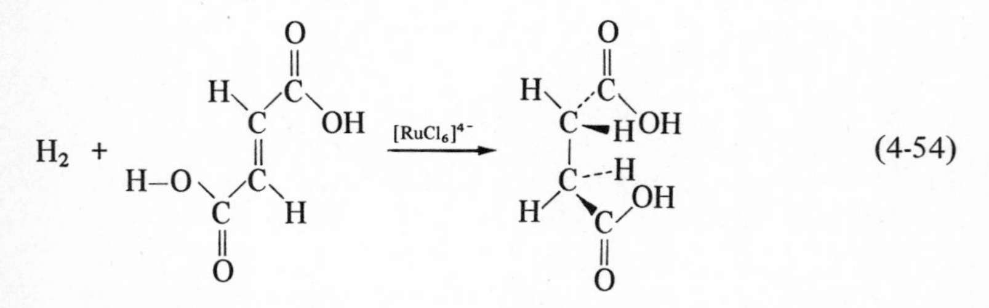

The detailed mechanism of the reaction is given in Eq. (4-55); the six (octahedral) ligand positions are occupied by chloride ions (not shown) which are displaced when other ligands enter and return when a ligand position becomes vacant.

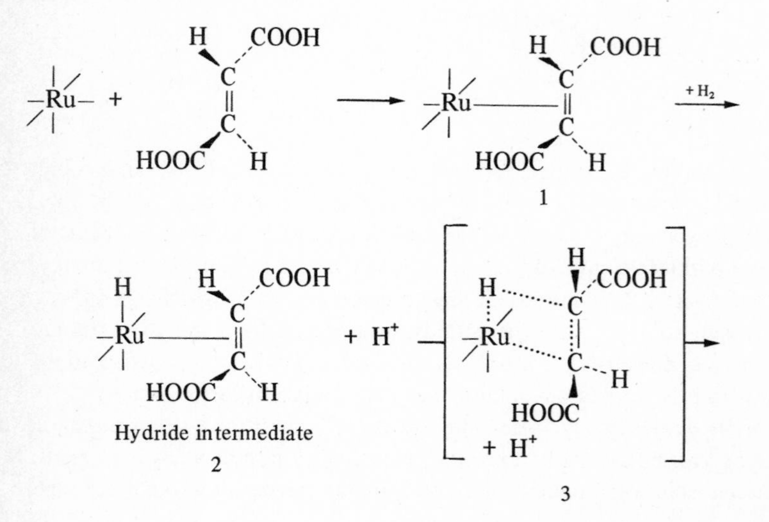

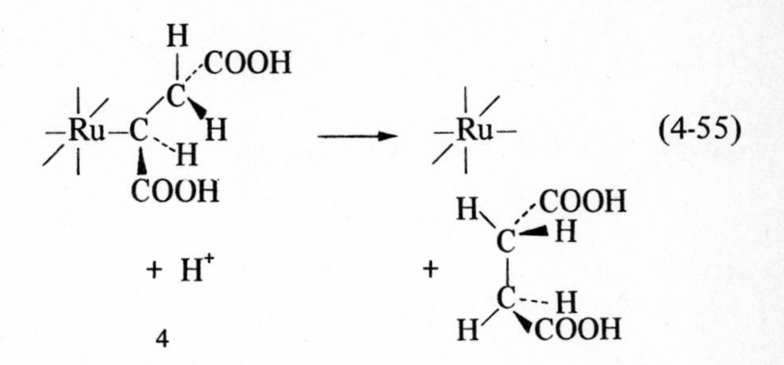

Alkenes are good ligands and the π complex, 1, forms readily. Its further reaction with dissolved hydrogen to form the hydride intermediate, 2, is the slow step of the reaction. Consequently, the hydride intermediate is never at a high enough concentration to be detected, but it is assumed to exist because a similar ruthenium(II) hydride complex, in which some of the chloride ligands are replaced by other ligands, is known to exist. The reactions of species 2 via transition state 3 to give species 4 may be called a ligand migration reaction in which the hydride ligand migrates to the alkene ligand while the latter simultaneously alkylates ruthenium. Species 4 subsequently reacts with a proton to give the final hydrogenated product and a free catalyst species that can react with another substrate molecule.

Replacing hydrogen (H_2) by deuterium (D_2) gas, and using deuterium oxide (D_2O) as solvent in place of ordinary water, has revealed further details of this reaction. First, when deuterium gas (D_2) is used in place of hydrogen (H_2) and ordinary water (H_2O) is the solvent, *no* deuterium atoms are found in the succinic acid product. But the reaction of complex 1 with a deuterium molecule should give a deuteride intermediate corresponding to species 2 with a deuteride ligand in place of the hydride. Apparently, the deuteride (or hydride) ligand can react very rapidly with a solvent water molecule (H_2O), thus being replaced by a hydride ion; this reaction must be much faster than the reaction of species 2 to give species 4, otherwise deuterium would appear in the product, contrary to what is observed.

A second observation, when D_2O is the solvent, has to do with the stereochemistry of the reaction. The preceding implies, and experiment agrees, that when D_2O is used as solvent instead of H_2O, two deuterium atoms add to the double bond of the substrate. The result is that the product contains two asymmetric carbon atoms whose configurations yield information about the stereochemistry of the reaction [Eq. (4-56)].

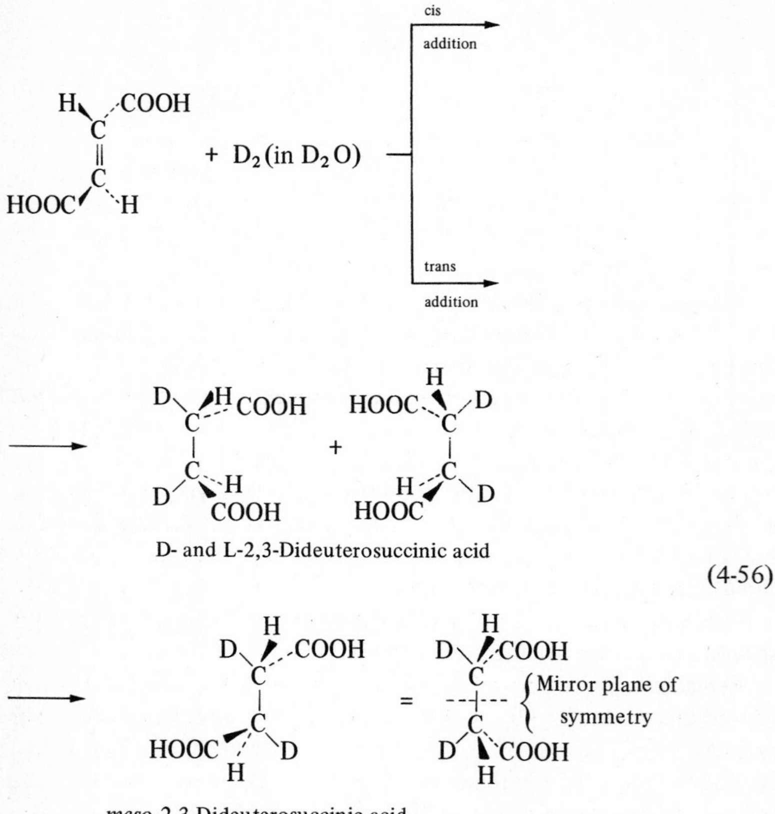

If both deuterium atoms add from the same side of the double bond (cis addition), the two asymmetric atoms in any given product molecule will both have either the D or L configuration. If the deuterium atoms add from opposite sides of the double bond (trans addition), the product is a different stereoisomer called the meso form in which one asymmetric carbon has the D and the other the L configuration. The infrared absorption spectrum of the actual products of the experiment show that D and L isomers, but not meso isomers, are formed. Hence cis addition occurs. Since the first atom added comes from the ruthenium side of the substrate [Eq. (4-55)], the implication of cis addition is that the proton attacks species 4 from the ruthenium side also rather than from the opposite side.

Homogeneous catalysts for alkene hydrogenation, such as the ruthenium(II) compound just described, have been discovered only during the last decade. On the other hand, although heterogeneous catalysts for the same reactions have been known since before the turn of the century, their mechanisms of action are only now being ascertained. In fact one of the main pieces of support for the current view that heterogeneous hydrogenation catalysts also form intermediate hydride compounds between hydrogen atoms and surface metal atoms is the argument by analogy, namely, that such compounds are believed to exist in homogeneous catalyses of the type we have just described. The mechanisms of hydrogenation of ethylene (gas) catalyzed by chromia gel (solid) is shown in Eq. (4-57).

$$\begin{array}{c} \text{H–H(gas)} \\ + \\ \underbrace{\text{–O–Cr–O–}}_{\text{At catalyst surface}} \end{array} \longrightarrow \begin{array}{c} \text{H}\quad\text{H} \\ |\quad\;\; | \\ \underbrace{\text{–O–Cr–O–}} \end{array} \xrightarrow{H_2C{=}CH_2}$$

$$\begin{array}{c} CH_3 \\ | \\ CH_2\text{H} \\ |\quad\; | \\ \underbrace{\text{–O–Cr–O–}} \end{array} \longrightarrow \begin{array}{c} H_3C\text{–}CH_3 \\ + \\ \underbrace{\text{–O–Cr–O–}} \end{array} \qquad (4\text{-}57)$$

It is of interest to note that heterogeneous hydrogenation catalysts have been used commercially for many years (e.g., for the hydrogenation of vegetable oils to give margarine) even though the mechanism of the catalysis was not understood.

Cyclooligomerization of Acetylene

Certain nickel compounds catalyze the formation of cyclooctatetraene from acetylene [Eq. (4-58)]. The reaction proceeds homogeneously in hydrocarbon solvents at 90°C and 20 atm of acetylene.

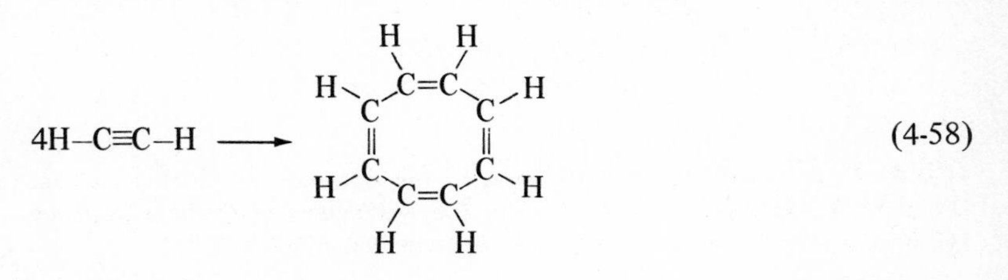

Octahedral nickel(II) complexes containing two ligands, such as cyanide ions, are effective catalysts. The remaining four ligand positions become occupied by acetylene molecules, and the reaction proceeds to product in a single step, as shown in Fig. 4-12*a*. This is a particularly good example of how a metal acts both as an activator for the reaction, and

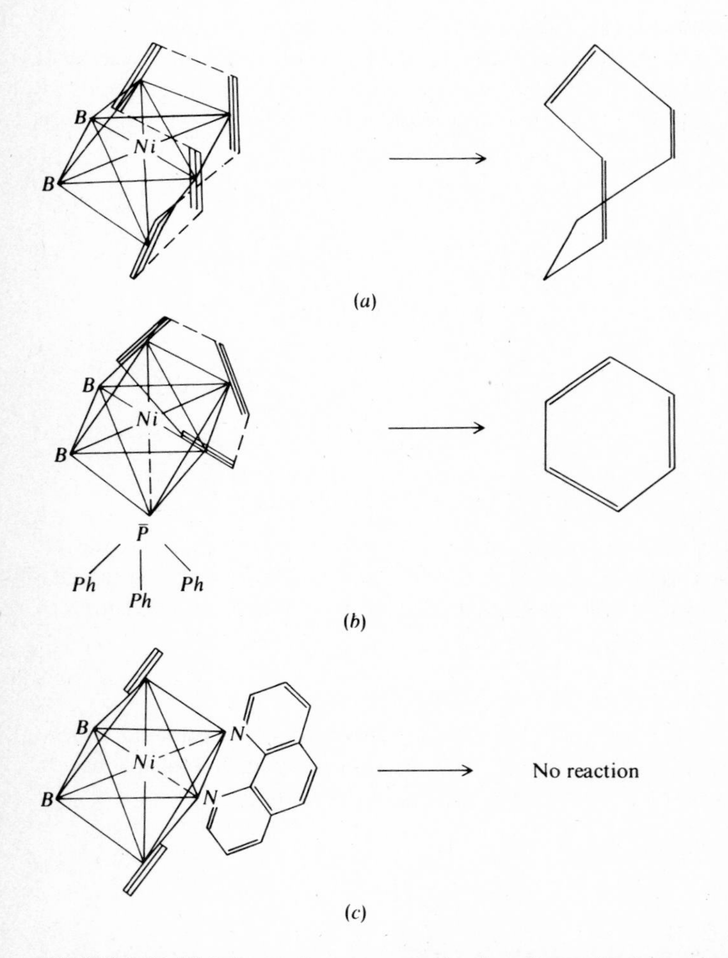

FIGURE 4-12 Nickel complex structures in the oligomerization of acetylene. [*From G. N. Schrauzer and S. Eichler, The Mechanism of Cyclooctatetraene Synthesis by Reppe's Method, Chemische Berichte, 95, 550 (1962).*]

as a template to bring the four reactant molecules together in the correct orientation.

If another good ligand, such as triphenylphosphine, is added to the solution in a 1:1 molar ratio relative to the nickel complex, only three acetylene molecules can be complexed. As one would predict, the observed product in this case is benzene (Fig. 4-12*b*). Finally, if *o*-phenanthroline (a bidentate ligand) is added to the solution, a square-planar complex of nickel forms (Fig. 4-12*c*). The two acetylene molecules that can bind in this case should be too far apart to react; indeed, under these conditions, no reaction is observed.

Vitamin B_{12}

One of the chemical forms of the coenzyme vitamin B_{12} is shown in Fig. 4-13. Its biochemical role is not yet well understood, but the cobalt atom, when reduced to the 1+ oxidation state, is extremely nucleophilic and easily alkylated. In this reaction the alkyl group displaces the cyanide ligand giving one of the few naturally occurring organometallic compounds. It appears that vitamin B_{12} may act as an intermediary in alkyl group transfers such as the transfer of a methyl group from another coenzyme, the methyltetrahydrofolate ion (CH_3–THF), to carbon dioxide to give acetic acid [Eq. (4-59)].

$$CH_3\text{–THF} + Co(B_{12}) \longrightarrow \underset{+\ \text{THF}}{CH_3\text{–}Co(B_{12})} \xrightarrow{+CO_2} CH_3\text{–}\overset{\overset{\displaystyle O}{\|}}{C}\text{–}O^- + Co(B_{12}) \qquad (4\text{-}59)$$

In this compound the cobalt atom is in the 1+ oxidation state. The rather complex ligands in vitamin B_{12} have the property both of stabilizing this oxidation state and of facilitating both the formation and cleavage of the methyl-cobalt bond in Eq. (4-59).

Fixation of Nitrogen

Finally, an extremely exciting and important area of research involving metal ion catalysts is the search for a catalyst for the fixation of nitrogen, the conversion of molecular nitrogen (N_2) to ammonia or some other nitrogen compound. It seems unlikely that homogeneous catalyst systems will replace the efficient Haber process which utilizes a

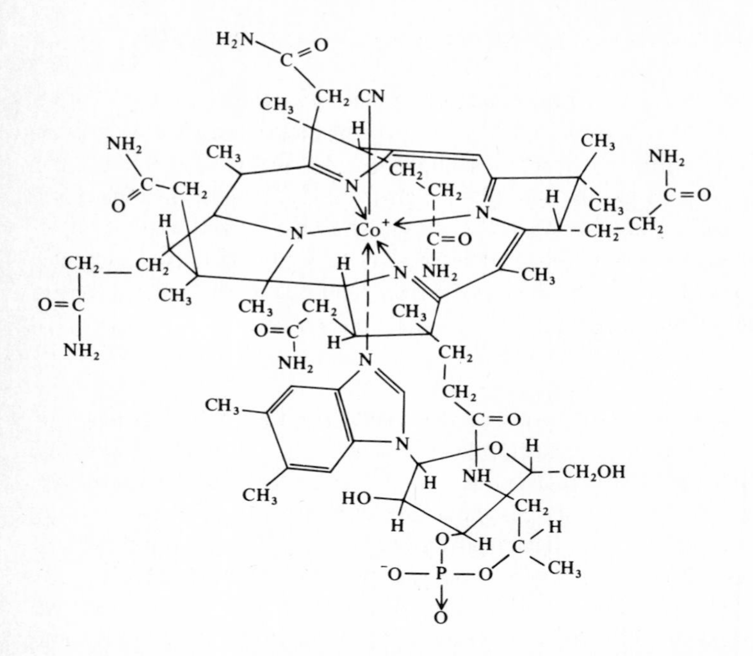

FIGURE 4-13 Structure of vitamin B_{12}. The cobalt ion is octahedrally coordinated. The complicated ring system with its four nitrogen ligands forms a plane. Another nitrogen ligand coordinates from below this plane and a cyanide ion from above. (*From M. Florkin and E. H. Stotz (eds.), "Comprehensive Biochemistry," vol. 11, p. 105, Elsevier Publishing Company, New York, 1963.*)

heterogeneous catalyst system. However, research in this area is being pursued in the hope of finding a catalyst that might enhance the efficiency of biological nitrogen fixation by the enzyme nitrogenase, thus decreasing the need for nitrogen fertilizers while improving crop yields.

In 1965 the first metal ion complexes containing molecular nitrogen as a ligand, such as the ruthenium(II) complex in Fig. 4-14*a*, were prepared. Since then dozens of such complexes have been made. One of the first "dimeric" complexes, in which the nitrogen ligand acts as a bridge between two metal ions, is shown in Fig. 4-14*b*.

More recently, dimeric complexes have been prepared in which one (or both) of the metal ions is (are) either molybdenum or iron. This is of particular interest because iron and molybdenum are present in the enzyme nitrogenase (found in legume plants and microorganisms),

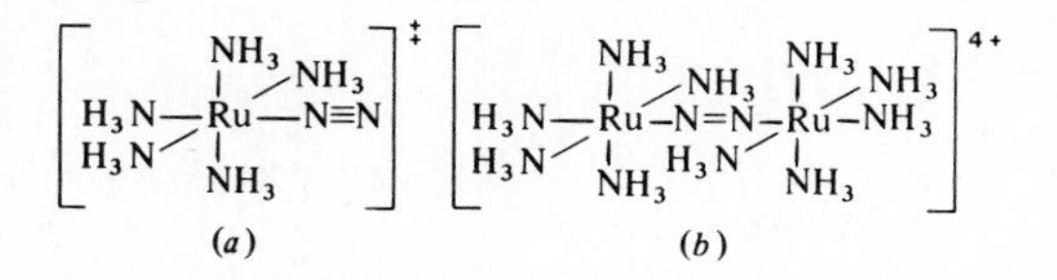

FIGURE 4-14 Complexes with molecular nitrogen ligands.

which catalyzes the biological reduction of molecular nitrogen to ammonia. It has been suggested that the metal ions in nitrogenase might also contain hydride ligands which could react with a complexed nitrogen molecule [Eq. (4-60)] in one of the early steps of the enzymatic reaction. This suggestion, although reasonable, is speculative at this point.

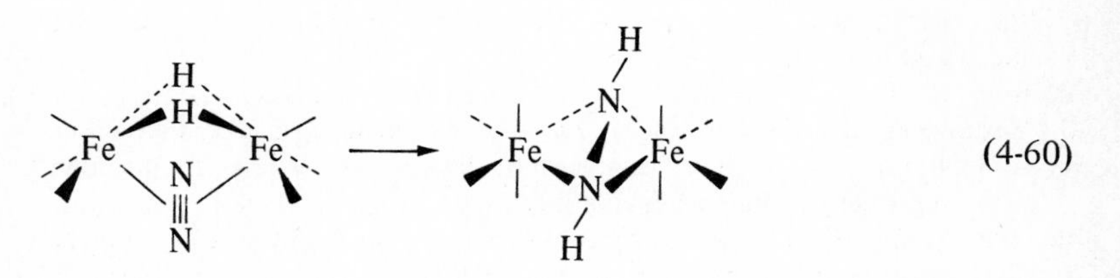

(4-60)

Hopefully, the current research into nitrogen-binding metal ions will lead to the discovery of new complexes in which the nitrogen ligand is sufficiently activated so that it will readily react with mild reducing agents to give ammonia. Four requirements of such a catalyst would seem to be the ability (1) to accept gaseous nitrogen as a ligand, (2) to form metal hydrides by oxidative addition of hydrogen, (3) to promote hydride migration to coordinated nitrogen, and (4) to dissociate the coordinated ammonia so produced.

Some metal ion systems have been reported to truly catalyze the reduction of molecular nitrogen to ammonia. However, these systems rapidly lose their catalytic ability and they require very strong reducing agents. Few details of the mechanism have been elucidated. Nevertheless, in view of the rapid progress in this field and the considerable effort being expended to develop metal ion catalyst systems and to elucidate the mechanism of nitrogenase action, it is perhaps not too optimistic to expect that a commercially useful metal ion fixation catalyst will be found in the not-too-distant future.

SUGGESTED READINGS

Williams, R. J. P.: Role of Transition Metal Ions in Biological Processes, *Royal Institute of Chemistry Reviews*, vol. 1, p. 13, 1963.

Cotton, F. A.: Ligand Field Theory, *Journal of Chemical Education*, vol. 41, p. 466, 1964. Critique of theories on the nature of bonding in metal ion complexes.

Westheimer, F. H.: The Mechanisms of Some Metal Ion Promoted Reactions, *Transactions of the New York Academy of Sciences*, vol. 18, p. 15, 1955.

Klotz, I. M.: in W. D. McElroy and B. Glass (eds.), "The Mechanism of Enzyme Action," p. 275, Johns Hopkins Press, Baltimore, Md., 1954.

Hamilton, G. A.: Oxidation by Molecular Oxygen, *Journal of the American Chemical Society*, vol. 86, p. 3391, 1964.

Wilke, G.: *Proceedings of the Robert A. Welch Foundation Conference on Chemical Research*, vol. 9, p. 165, 1965.

Natta, G.: Precisely Constructed Polymers, *Scientific American*, vol. 205, p. 33, August, 1961. An excellent description of stereoregular polymers by one of the scientists who opened up this field and still contributes to it. Very well illustrated.

Burwell, R. L.: The Mechanism of Heterogeneous Catalysis, *Chemical and Engineering News.*, p. 56, August 22, 1966. Good survey of the mechanisms of hydrogenation reactions catalyzed by heterogeneous catalysts including some experimental approaches currently used.

Halpern, J.: Coordination Chemistry and Homogeneous Catalysis, *Chemical and Engineering News*, p. 68, October 31, 1966. Deals with the mechanisms of certain reactions catalyzed by metal ion complexes in homogeneous solution and some of the general principles that underlie this field.

Haensel, V., and R. L. Burwell, Jr.: Catalysis, *Scientific American*, vol. 225, p. 46, December, 1971. A discussion of heterogeneous catalyst systems with emphasis on commercially important processes.

Gilman, H., and J. J. Eisch: Lithium, *Scientific American*, vol. 208, p. 88, January, 1963. The unique ability of lithium to catalyze polymerization reactions is discussed.

Postgate, J.: "Chemistry and Biochemistry of Nitrogen Fixation," Plenum Press, London, 1971.

Schneller, S. W.: Nitrogen Fixation, an Interdisciplinary Frontier, *Journal of Chemical Education*, vol. 49, p. 786, 1972. A brief review of the state of knowledge of nitrogenase action and of metal ion complexes related to nitrogenase.

FIVE
CATALYSIS BY NUCLEOPHILES AND ELECTROPHILES

In Chapter 3 we discussed nucleophile catalysis in the hydrolysis of carboxylic acid derivatives. We now want to discuss some other selected examples of nucleophile- and electrophile-catalyzed reactions, involving catalysts and types of reactions which are closely related to biological systems. As we shall see, many of the most important coenzymes act as nucleophile or electrophile catalysts.

Nucleophiles are molecules or anions in which one atom has an unshared pair of electrons which it can readily share with an electron-deficient atom in another molecule to form a covalent bond. The latter molecule is defined as an electrophile; it has an atom capable of accepting a share in a pair of electrons. (The words "nucleophile" and "electrophile" are derived from the Greek and mean literally nucleus-loving and electron-loving, respectively.) Thus a nucleophile N reacts with an electrophile E or vice versa [first step in Eq. (5-1)]. If, as a result of this combination, either E or N undergoes a reaction more readily than it normally would, we have either nucleophile or electrophile catalysis, respectively.

$$\mathrm{E} + {:}\mathrm{N} \longrightarrow \mathrm{E{-}N} \begin{cases} \xrightarrow{\text{nucleophile catalysis}} \text{product} + {:}\mathrm{N} \\ \xrightarrow{\text{electrophile catalysis}} \text{product} + \mathrm{E} \end{cases} \tag{5-1}$$

The essential feature of a nucleophile or electrophile catalyst is that it reacts with the substrate to form an intermediate which then reacts further to give the final products and regenerate the catalyst. This is illustrated by the reaction that we discussed in Sec. 1-3, the hydrolysis of methyl iodide which is subject to nucleophile catalysis by the bromide ion. In general a catalyst opens up a new pathway in which all steps are faster than the slowest step in the uncatalyzed pathway.

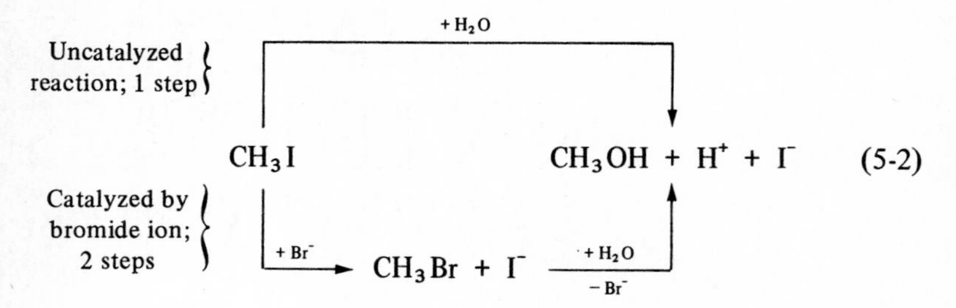

The bromide ion is a catalyst because both steps in the catalyzed pathway of Eq. (5-2) are faster than the single step in the uncatalyzed reaction. Specifically this means that (1) the bromide ion displaces the iodide ion more readily than water does, i.e., the bromide ion is a better nucleophile than water; and (2) methyl bromide reacts with water more readily than does methyl iodide, i.e., the bromide ion is a better leaving group than the iodide ion in this reaction.

In general a good nucleophile catalyst is a substance which is a relatively good nucleophile (so it reacts readily with the substrate) and a relatively good leaving group. Very often these two characteristics are antagonistic, i.e., the best nucleophiles are poor leaving groups and conversely the best leaving groups are poor nucleophiles; thus, a very good nucleophile is not necessarily a good catalyst. However, some nucleophiles, because of their electronic makeup, are able to combine relatively good nucleophilicity with relatively good leaving ability and hence are good catalysts. In the following sections we shall point out some of the factors contributing to this.

What we have just said about nucleophile catalysts holds true also for electrophile catalysts, as we shall see later in this chapter. One can substitute the word electrophile for nucleophile in most of the preceding general statements. Thus, a good electrophile catalyst is one which is a relatively good electrophile and a relatively good leaving group.

5-1 CATALYSIS BY NUCLEOPHILES

Cyanide Ion Catalysis of Acetoin Formation

When cyanide ion (e.g., in the form of sodium cyanide) is added to a solution of acetaldehyde, the latter undergoes a dimerization reaction giving acetoin [Eq. (5-3)]. The detailed mechanism for the reaction is given in Eqs. (5-4) and (5-5).

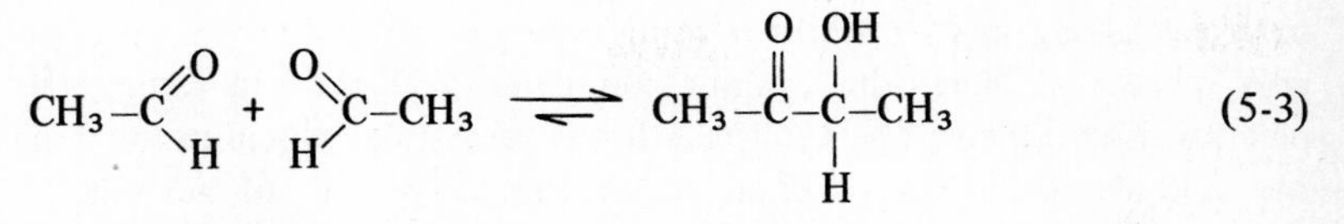

$$\mathrm{CH_3{-}C({=}O){-}H} + {}^{-}\mathrm{C{\equiv}N} \rightleftharpoons \left[\mathrm{CH_3{-}C(\overset{\cdots}{=}O)({\cdots}C{\equiv}N){-}H}\right]^{-} \rightleftharpoons \underset{1}{\mathrm{CH_3{-}C(O^{-})(C{\equiv}N){-}H}} \rightleftharpoons \underset{2}{\mathrm{CH_3{-}C^{-}(OH){-}C{\equiv}N}} \quad (5\text{-}4)$$

$$\mathrm{CH_3{-}C^{-}(OH){-}C{\equiv}N} + \mathrm{H{-}C({=}O){-}CH_3} \rightleftharpoons \left[\mathrm{CH_3{-}C(OH)(C{\equiv}N){\cdots\cdots}C(\overset{\cdots}{=}O)(H){-}CH_3}\right]^{-} \rightleftharpoons$$

$$\underset{3}{\mathrm{CH_3{-}C(OH)(C{\equiv}N){-}C(O^{-})(H){-}CH_3}} \rightleftharpoons$$

$$\underset{4}{\mathrm{CH_3{-}C(^{-}O)(C{\equiv}N){-}C(HO)(H){-}CH_3}} \rightleftharpoons \left[\mathrm{CH_3{-}C(\overset{\cdots}{=}O)({\cdots}C{\equiv}N){-}C(HO)(H){-}CH_3}\right]^{-} \rightleftharpoons \mathrm{CH_3{-}C({=}O){-}C(OH)(H){-}CH_3} + {}^{-}\mathrm{C{\equiv}N} \quad (5\text{-}5)$$

The cyanide ion first adds to the carbonyl carbon atom of acetaldehyde to give species 1 which rapidly undergoes a proton shift to give the carbanion, species 2. The latter then adds to a second acetaldehyde molecule giving species 3, which again undergoes a rapid proton shift to give species 4. When the cyanide ion departs in the final step, the reaction is complete. The total reaction is reversible, which means that the cyanide ion is also a catalyst for the conversion of acetoin to acetaldehyde. However, the equilibrium constant for Eq. (5-3) is large; therefore, the reaction goes essentially completely to the acetoin form.

The cyanide ion is a particularly effective catalyst for this reaction; very few other nucleophiles are. Why? Normally a carbanion intermediate has a very high free energy and is difficult to form. However, in species 2 the negative charge is not totally localized on one carbon atom, as shown in Eq. (5-4), but is partly shared by the cyano group. One could equally well write species 2 with the negative charge localized on the nitrogen atom, as shown in Eq. (5-6) (species 2*a*).

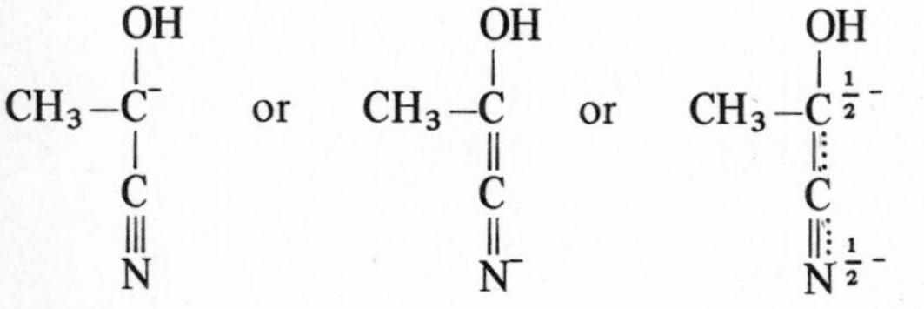

$$\underset{2}{CH_3-\overset{\overset{\displaystyle OH}{|}}{\underset{\underset{\underset{\displaystyle N}{|||}}{\overset{|}{C}}}{C^-}}} \quad \text{or} \quad \underset{2a}{CH_3-\overset{\overset{\displaystyle OH}{|}}{\underset{\underset{\underset{\displaystyle N^-}{||}}{\overset{||}{C}}}{C}}} \quad \text{or} \quad \underset{2b}{CH_3-\overset{\overset{\displaystyle OH}{|}}{\underset{\underset{\underset{\displaystyle N^{\frac{1}{2}-}}{|||}}{\overset{|:}{C}}}{C^{\frac{1}{2}-}}}} \qquad (5\text{-}6)$$

In fact the actual electron distribution in the carbanion is somewhat intermediate between these two extreme representations, probably something like species 2*b* in Eq. (5-6). At any rate, sharing the negative charge over several atoms greatly lowers the free energy of species 2. Furthermore, in the cyanide ion itself, the negative charge is about equally shared by both atoms. As a leaving group, the cyanide ion carries a negative charge with it. Since this charge can be shared by two atoms, less energy is required to create this negative charge than if it were localized on one atom. Thus, the cyanide ion is a good leaving group.

Thiamin Catalysis

The coenzyme thiamine pyrophosphate (Fig. 5-1*a*) has catalytic capabilities very similar to those of the cyanide ion. Even in the absence of enzyme it is a catalyst, like the cyanide ion, for the formation of

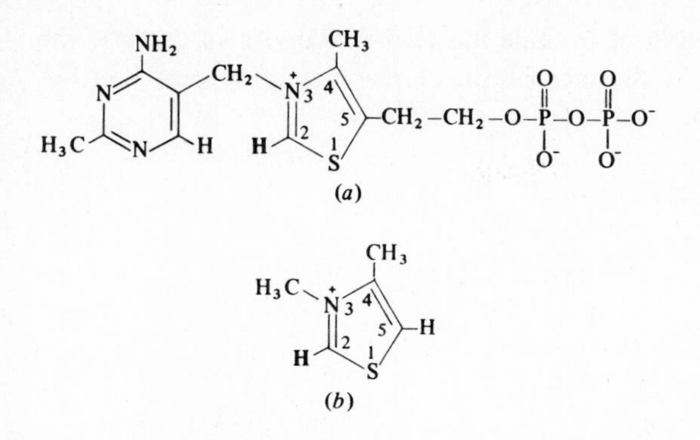

FIGURE 5-1 Structures of (*a*) thiamine pyrophosphate and (*b*) 3,4-dimethylthiazolium ion.

acetoin from acetaldehyde. The 3,4-dimethylthiazolium cation (Fig. 5-1*b*) also catalyzes this reaction. Thus the thiazole ring part of thiamine pyrophosphate is the catalytically active part of the molecule; the rest presumably is necessary to bind it to various enzymes for which it is a coenzyme. For simplicity we shall represent thiamine in the following discussion as a substituted thiazole ring [e.g., Eq. (5-7)].

When 3,4-dimethylthiazolium ion is dissolved in D_2O, the boldface hydrogen atom at carbon atom 2 (Fig. 5-1) of the thiazole ring is replaced by a deuterium atom. The reaction is essentially complete in less than 1 min. This implies that the proton readily dissociates from carbon atom 2 of the thiazole ring. The resulting carbanion is relatively stable because the negative charge is partly shared by the nitrogen atom, analogous to the cyanide ion [Eq. (5-7)].

Since the nitrogen atom in the thiazole ring was originally positively charged, it probably takes on an even greater share of the negative charge than in the cyanide ion. In the acetoin condensation this anion of thiamine pyrophosphate adds to acetaldehyde [Eq. (5-8)] in direct analogy to the reaction of cyanide ion [Eq. (5-4)]. As an exercise the reader may write out the mechanism of reaction of species 1 in Eq. (5-8) to give acetoin.

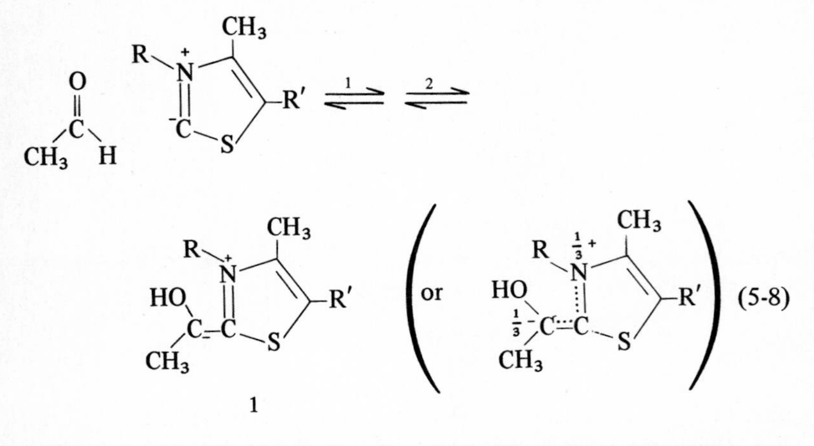

In enzyme-catalyzed reactions for which thiamine pyrophosphate is a coenzyme, there is good evidence that the latter functions in the same way we have just described for nonenzymic reactions. The enzyme pyruvate carboxylase together with thiamine pyrophosphate catalyzes the decarboxylation of pyruvic acid and other α-keto acids [Eq. (5-9)].

$$\underset{\text{Pyruvate ion}}{CH_3-\overset{\overset{\displaystyle O}{\|}}{C}-\overset{\overset{\displaystyle O}{\|}}{C}-O^-} \longrightarrow \underset{\text{Acetaldehyde}}{CH_3-\overset{\overset{\displaystyle O}{\|}}{C}-H} + CO_2 \qquad (5\text{-}9)$$

The detailed mechanism for this reaction [Eq. (5-10)] is essentially identical to the cyanide ion catalysis of Eq. (5-4), except that a carbon-carbon bond is broken in the rate-determining step instead of a carbon-hydrogen bond. Thiamine pyrophosphate alone will catalyze the reaction, but not as efficiently as it does in the presence of the enzyme. How the enzyme participates in the reaction is not understood, so only the coenzyme participation is shown. In support of the mechanism of Eq. (5-10) is the fact that species 1 has actually been isolated from solutions in which the enzyme-catalyzed decarboxylation is taking place.

Other reactions of pyruvate ion plus thiamine pyrophosphate besides route 11*a* can occur in the presence of the appropriate enzyme and other reactants. In all of these reactions intermediate 1 of Eq. (5-10) probably participates and gives different products depending on what other reagents are present in the solution. Thus, if an oxidizing agent is present, acetic acid is eventually produced (route 11*b*); if acetaldehyde is present, acetoin is produced (route 11*c*).

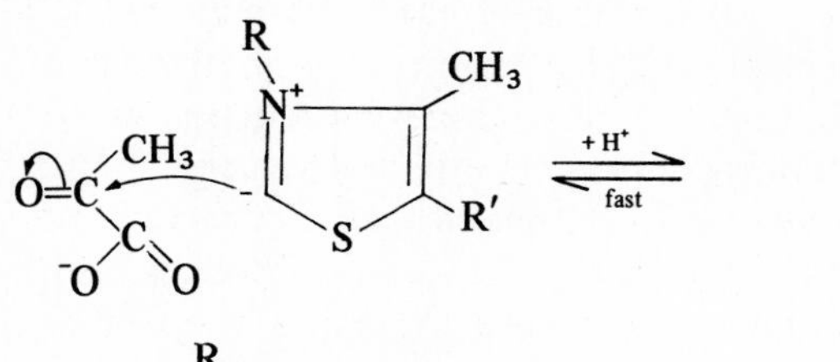

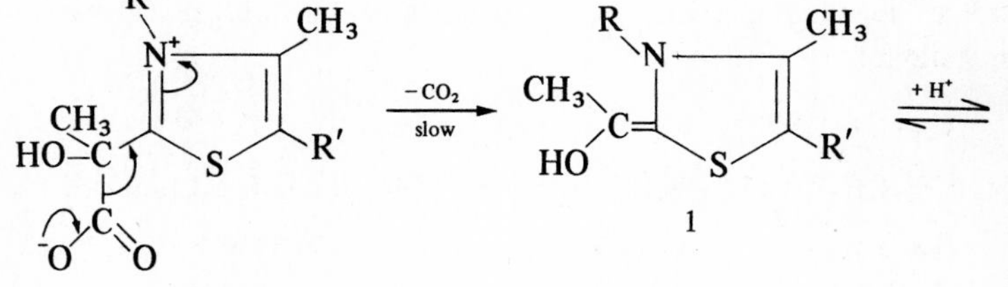

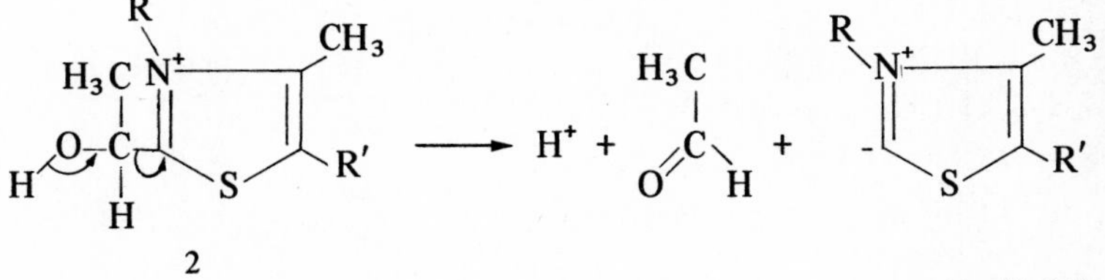

(5-10)

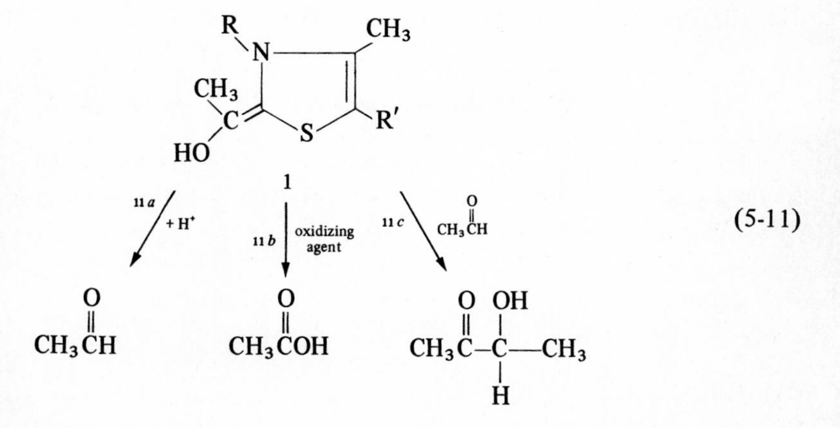

(5-11)

Thiols as Nucleophile Catalysts

In compounds of the general type RSH, the sulfur of the thiol group (–SH group) is a particularly good nucleophile. As a catalyst it is not particularly good, however, because it tends to be a rather poor leaving group. Nevertheless, in certain instances thiol compounds are of considerable biological importance. One of the most important coenzymes is coenzyme A in which a thiol group is the reactive part of the molecule. Among its many functions coenzyme A acts as an acyl carrier and activator by forming thiol esters with acyl groups (Fig. 5-2).

Coenzyme A (abbreviated HSCoA) is particularly important in the biological degradation of fatty acids, a series of reactions which provides considerable energy for the organism. The overall process involves many steps, but we shall break it up into three steps for convenience.

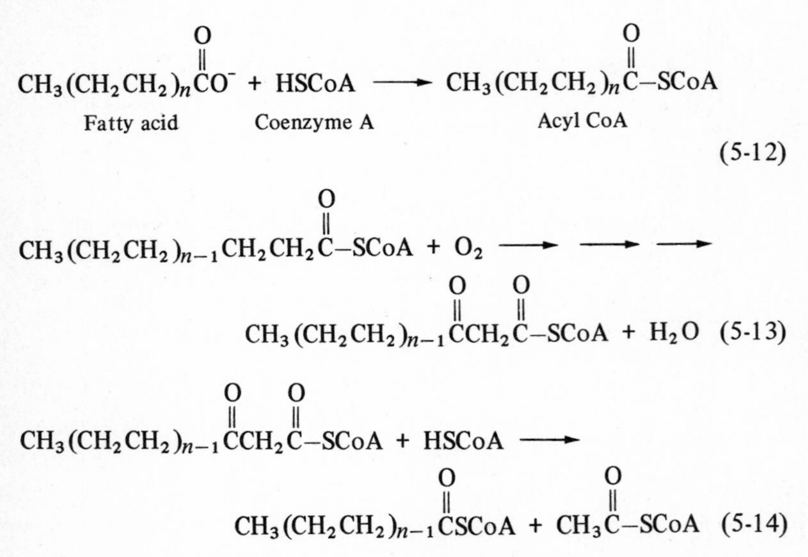

$$\underset{\text{Fatty acid}}{CH_3(CH_2CH_2)_n\overset{O}{\overset{\|}{C}}O^-} + \underset{\text{Coenzyme A}}{HSCoA} \longrightarrow \underset{\text{Acyl CoA}}{CH_3(CH_2CH_2)_n\overset{O}{\overset{\|}{C}}\text{–SCoA}} \tag{5-12}$$

$$CH_3(CH_2CH_2)_{n-1}CH_2CH_2\overset{O}{\overset{\|}{C}}\text{–SCoA} + O_2 \longrightarrow \longrightarrow \longrightarrow CH_3(CH_2CH_2)_{n-1}\overset{O}{\overset{\|}{C}}CH_2\overset{O}{\overset{\|}{C}}\text{–SCoA} + H_2O \tag{5-13}$$

$$CH_3(CH_2CH_2)_{n-1}\overset{O}{\overset{\|}{C}}CH_2\overset{O}{\overset{\|}{C}}\text{–SCoA} + HSCoA \longrightarrow CH_3(CH_2CH_2)_{n-1}\overset{O}{\overset{\|}{C}}SCoA + CH_3\overset{O}{\overset{\|}{C}}\text{–SCoA} \tag{5-14}$$

The first step [Eq. (5-12)] has an unfavorable equilibrium constant, and so the energy from the hydrolysis of an ATP molecule is required to

$$\underset{\text{Ester}}{R\text{–}\overset{O}{\overset{\|}{C}}\text{–O–}R'} \qquad \underset{\text{Thiol ester}}{R\text{–}\overset{O}{\overset{\|}{C}}\text{–S–}R'}$$

FIGURE 5-2

drive it to completion. The acylated coenzyme A now undergoes several oxidation reactions leading to the formation of a β-keto group in the acyl part of the molecule [Eq. (5-13)]. Finally in Eq. (5-14), another molecule of coenzyme A causes a cleavage of the acyl group to give a new acylated coenzyme A which can recycle repeatedly through Eqs. (5-13) and (5-14) until the original fatty acid has been completely converted into *n* molecules of acetylated coenzyme A (acetyl CoA).

Equation (5-14) illustrates the action of the thiol group of coenzyme A as a nucleophile. Although the mechanism of the enzyme-catalyzed process has not been determined, it is quite reasonable to assume that the reaction proceeds as shown in Eq. (5-15), probably with some additional steps involving the enzyme itself.

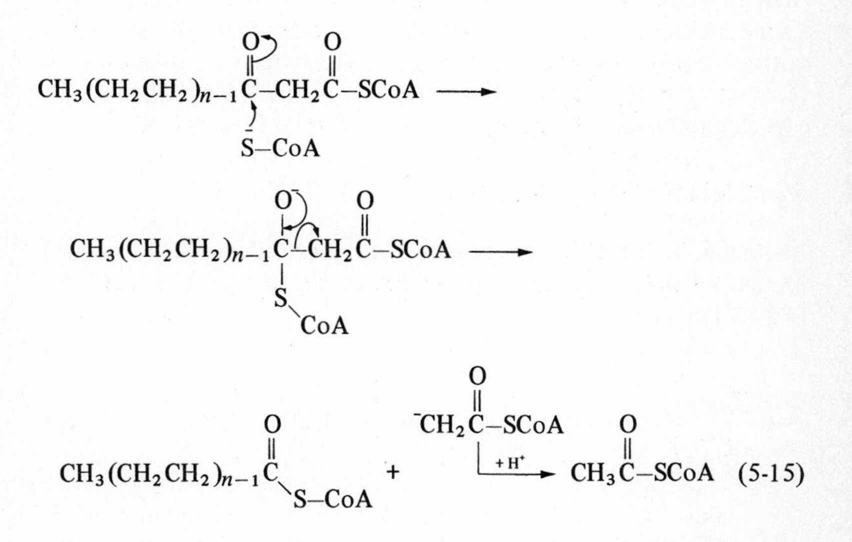

The further oxidations of the acetyl group in acetyl CoA via the citric acid cycle provides additional energy for the organism. Alternatively, acetyl CoA is, like ATP, a source of free energy [Eq. (5-16)]. Thus its hydrolysis can be used to drive other unfavorable reactions.

$$CH_3\overset{O}{\overset{\|}{C}}SCoA + H_2O \longrightarrow CH_3\overset{O}{\overset{\|}{C}}O^- + H^+ + HSCoA$$

$$\Delta G^\circ_{pH\,7} = -7.7 \text{ kcal mol}^{-1} \qquad (5\text{-}16)$$

For example, in the presence of the appropriate enzymes, the hydrolysis of acetyl CoA can form a molecule of ATP from ADP and phosphoric acid.

Other thiol-containing coenzymes include glutathione (glu.cys.gly) and lipoic acid. These sometimes function as redox catalysts as well as nucleophiles because the half-reaction of Eq. (5-17) occurs readily in either direction.

$$2R{-}SH \rightleftharpoons R{-}S{-}S{-}R + 2H^+ + 2e^- \qquad (5\text{-}17)$$

In a large number of enzymes, the thiol groups of cysteine residues may act as nucleophile catalysts. An example is papain which, like chymotrypsin, catalyzes the hydrolysis of esters, amides, and proteins. Its mechanism of action bears similarities to that of chymotrypsin. For instance, papain forms an intermediate acyl-enzyme with its substrates. However, the acyl-papain intermediate is a thiol ester, whereas the acyl-chymotrypsin intermediate is an oxygen ester (Sec. 3-12).

5-2 CATALYSIS VIA IMINE INTERMEDIATES

An imine, or a Schiff base as it is sometimes called, is a compound formed by the condensation of an amine with a ketone or aldehyde [Eq. (5-18)].

$$O{=}C(R)(R') + H_2\ddot{N}{-}R'' \rightleftharpoons (R)(R')C{=}N{-}R'' + H_2O \qquad (5\text{-}18)$$

The amine is a nucleophile and the carbonyl compound is an electrophile. If the carbonyl compound undergoes a reaction which occurs more readily when it is converted to an imine, then the amine is acting as a nucleophile catalyst. Conversely, if the amine undergoes a reaction which occurs more readily when it is in the imine form, then the carbonyl compound plays the role of an electrophile catalyst.

We shall discuss examples of both nucleophile and electrophile catalysis via imine-type compounds. First, however, we must briefly consider the mechanism of imine formation and of its reverse, imine hydrolysis.

Imine formation occurs in two steps—(1) the formation of a carbinolamine and (2) its dehydration. (A carbinolamine is a compound in which an amine group and an alcohol group are both attached to the same carbon atom.) Both steps are catalyzed by general acids, and in addition, the second step is sometimes catalyzed by general bases. Thus the mechanism of imine formation is as shown in Eqs. (5-19) and (5-20), where BH and B: represent a general acid and a general base, respectively.

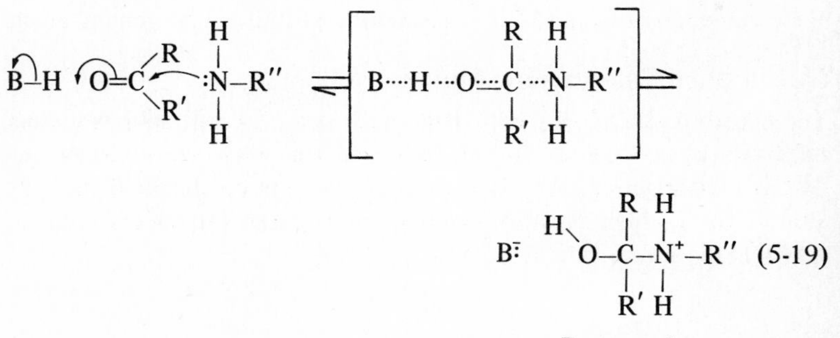

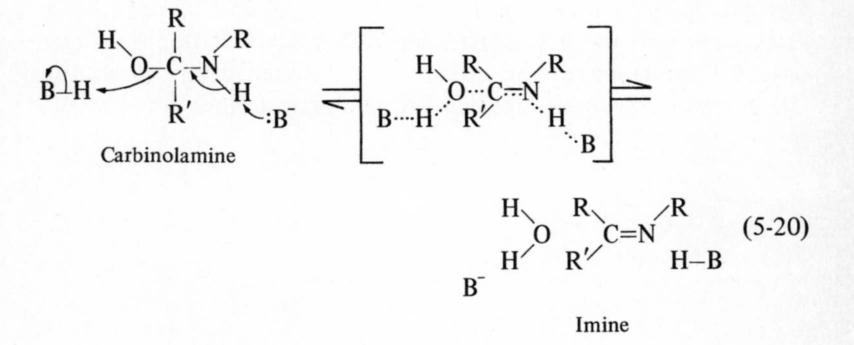

In the absence of other general acids and bases, water molecules will play the same role. We should also point out that the protonated carbinolamine formed in Eq. (5-19) will rapidly lose a proton, a step which has not been shown.

Now we turn to the reverse of imine formation, namely, imine hydrolysis. An important general principle is the *principle of microscopic reversibility* which states: the mechanism of any reaction in the backward direction is just the reverse of the mechanism in the forward direction. That is, if the mechanism is known for a reaction A → B → C, then the mechanism for the backward reaction, C → B → A, is just the reverse of the known mechanism. This is true no matter how many steps are involved in the mechanism. Thus imine hydrolysis occurs via the reverse of Eqs. (5-20) and (5-19). This implies further that where general acid catalysis occurs in the forward reaction, general base catalysis occurs in the reverse reaction, and vice versa. Thus in imine hydrolysis, both steps are catalyzed by general bases, and the first step [i.e., the reverse of Eq. (5-20)] is sometimes catalyzed by general acids.

Nucleophile Catalysis by Amines

The decarboxylation of β-keto acids, such as acetoacetic acid, is subject to catalysis by amines via imine formation. First we consider the uncatalyzed reaction. At pH 7 in aqueous solution this acid is in the anionic form which decarboxylates spontaneously, but very slowly, to give acetone and carbon dioxide [Eq. (5-21)].

$$CH_3\overset{\overset{O}{\|}}{C}CH_2\overset{\overset{O}{\|}}{C}O^- + H^+ \longrightarrow CH_3\overset{\overset{O}{\|}}{C}CH_3 + CO_2 \qquad (5\text{-}21)$$

The half-life of this reaction is about 23 days at 25°C. The probable mechanism is shown in Eqs. (5-22) and (5-23). The slow step is the first one, the formation of the intermediate 2 (often called an enolate ion–see Sec. 4-2) which accepts a proton to give acetone [Eq. (5-23)].

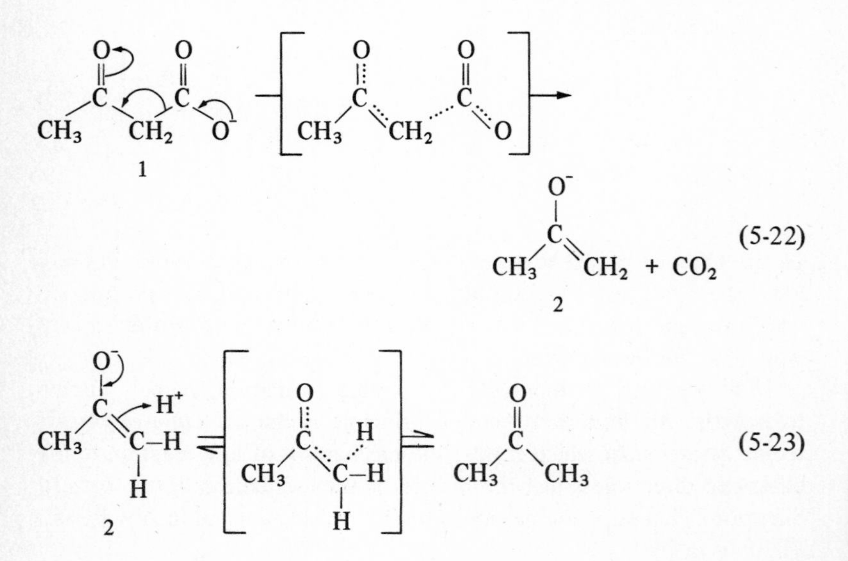

In acid solution, acetoacetic acid is in the undissociated form and the mechanism of the decarboxylation reaction [Eq. (5-24)] is slightly different.

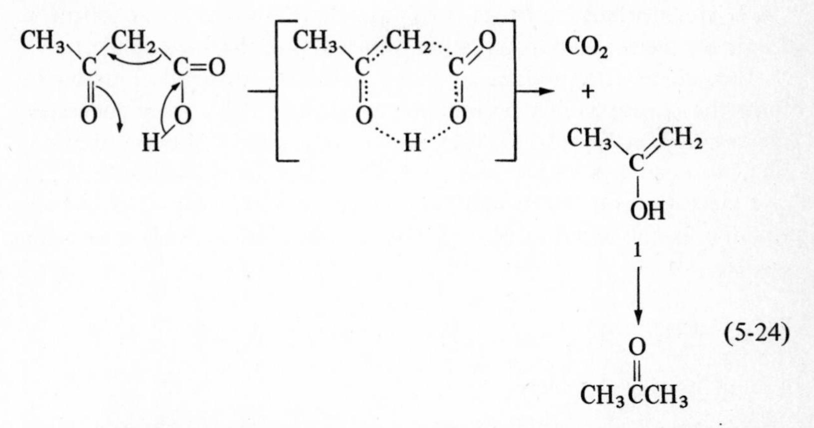

Here, the acidic hydrogen atom acts as an internal acid catalyst and assists the formation of species 1 in Eq. (5-24). The latter then undergoes a proton shift (tautomerization) to give acetone. The rate-determining step in Eq. (5-24) is about 50 times faster than in Eq. (5-22). This mechanism is reminiscent of metal ion-catalyzed decarboxylations which we discussed in Chap. 4. Here an internal proton catalyzes the reaction instead of a metal ion. Metal ions do not catalyze this reaction, presumably because the β-keto group by itself is a poor ligand for metal ions.

The decarboxylation of acetoacetate ion is catalyzed by primary amines, i.e., amines with only one hydrocarbon chain attached to the nitrogen atom. One particularly good amine catalyst is aminoacetonitrile which has an unusually low pK_a of 5.3 [Eq. (5-25)].

$$N{\equiv}C{-}CH_2{-}\overset{H}{\underset{H}{\overset{|}{\underset{|}{N^+}}}}{-}H + H_2O \rightleftharpoons$$

Acid form of aminoacetonitrile

$$N{\equiv}C{-}CH_2{-}\ddot{N}H_2 + H_3O^+ \qquad K_a = 5 \times 10^{-6} \quad pK_a = 5.3 \qquad (5\text{-}25)$$

Base form of aminoacetonitrile

Consequently, in neutral solution (pH 7), the amine is almost entirely in the base form, the only form which can act as a nucleophile; the acid (protonated) form has no unshared electron pair on the amino nitrogen atom. (Incidentally, there is no doubt that it is the amino nitrogen atom, not the cyano nitrogen atom, which acts as the nucleophile.)

The amine catalyst first combines with the acetoacetate anion to form the corresponding protonated imine. The latter then undergoes the reaction in Eq. (5-26), loss of CO_2 followed by proton addition. The protonated imine, product 3 in Eq. (5-26), readily hydrolyzes to give acetone and regenerate the amine catalyst. Thus the overall reaction is the same as Eq. (5-21) and the amine catalyst remains unchanged.

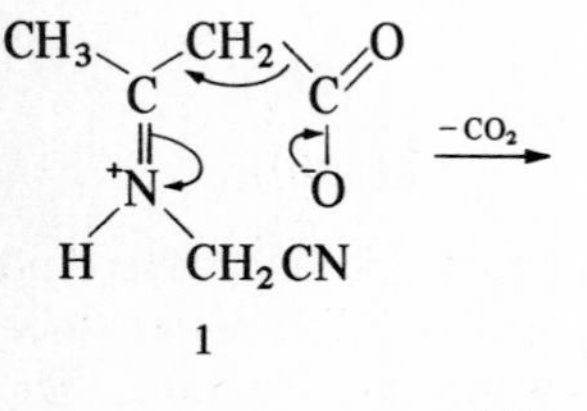

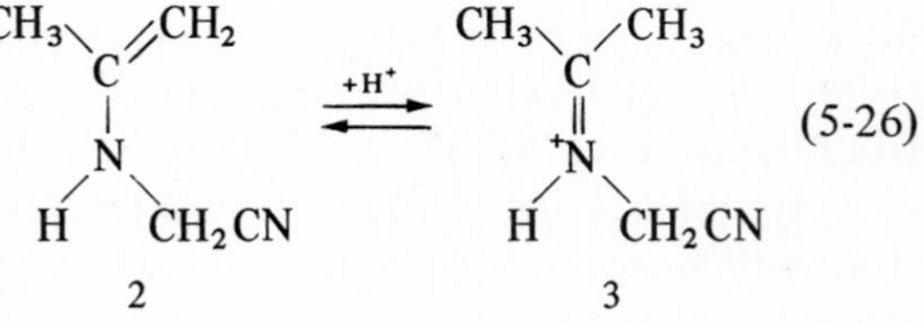

(5-26)

The loss of carbon dioxide in Eq. (5-26) is probably the rate-determining step in the reaction catalyzed by aminoacetonitrile just as in the uncatalyzed reaction [Eq. (5-22)]. Thus the role of the amine catalyst is to increase the rate of breaking the carbon-carbon bond. To see why it can do so, let us compare species 1 of Eq. (5-26) with species 1 of Eq. (5-22). The former has a positively charged nitrogen atom which is more capable of accepting an electron pair than is the neutral oxygen atom in the latter. One commonly generalizes by saying that the positively charged nitrogen atom is a good "electron sink."

In living organisms the enzyme acetoacetate decarboxylase catalyzes the decarboxylation of the acetoacetate ion. It is a very efficient catalyst. If we compare two solutions, each at pH 6 with an acetoacetate ion concentration of 1×10^{-3} *M*, one of which also contains the enzyme acetoacetate decarboxylase at a concentration of

1×10^{-5} *M*, the decarboxylation rate in the solution containing enzyme is about one million times larger than that of the uncatalyzed reaction.

What factors contribute to this high catalytic efficiency of the enzyme? One important factor is amine catalysis via imine formation of the type we have just discussed. The amine group of a particular lysine residue in the enzyme plays the same role as that of aminoacetonitrile in Eq. (5-26). For convenience, the mechanism for the enzyme-catalyzed reaction is reproduced in Eqs. (5-27) to (5-30), where ENH_2 represents the enzyme with its catalytic amine group.

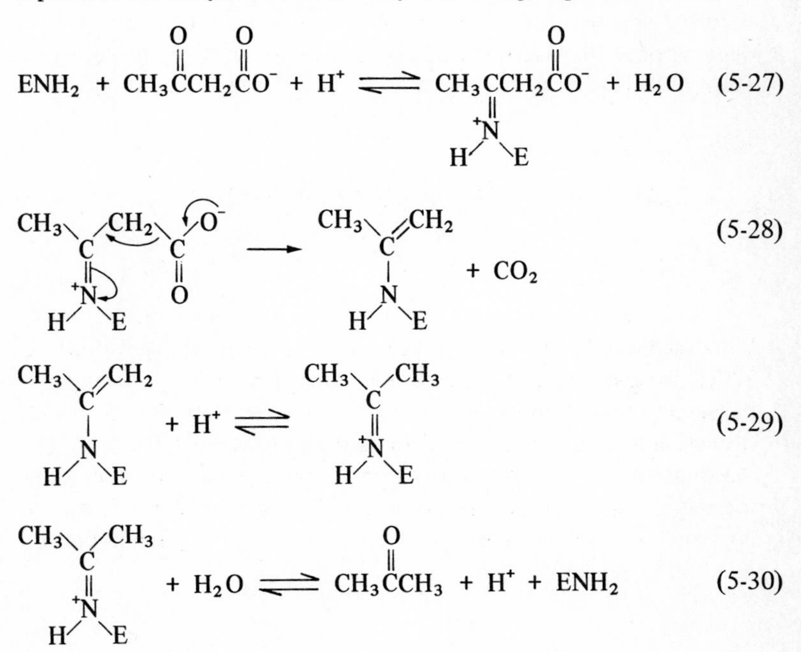

Which of these steps is rate-determining is not known; however, as before, it is quite possibly Eq. (5-28), the loss of carbon dioxide. The enzyme is a much more efficient catalyst than aminoacetonitrile although they have similar pK_a's; thus other factors besides simple amine catalysis must be involved. These include, no doubt, the decrease in entropy of activation through enzyme-substrate complex formation and the correct orientation of catalytic groups with respect to the substrate at the active site.

Although imine formation [Eq. (5-27)] and hydrolysis [Eq. (5-30)] are probably not rate-determining, they are probably also catalyzed by general acid and base groups at the active site of the enzyme. Without such catalysis, these steps could well be rate-determining and the enzyme would be less efficient than it is.

The following points of evidence for the mechanism of Eqs. (5-27) to (5-30) have been obtained by Westheimer at Harvard University.

1. In imine formation the carbonyl oxygen atom of the substrate is converted into a molecule of water. When the substrate which was labeled with oxygen-18 in the β-carbonyl group was added to an enzyme solution, the formation of $H_2{}^{18}O$ was observed, and it occurred much more rapidly than in the absence of enzyme. This demonstrates that an imine is formed between enzyme and substrate [Eq. (5-31)].

$$CH_3-\overset{\overset{^{18}O}{\|}}{\underset{\underset{H-N(H)-E}{|}}{C}}-CH_2\overset{\overset{O}{\|}}{C}O^- \longrightarrow CH_3-\underset{\underset{H\diagup\overset{+}{N}\diagdown E}{\|}}{C}-CH_2\overset{\overset{O}{\|}}{C}O^- + H_2{}^{18}O \qquad (5\text{-}31)$$

When acetone labeled with oxygen-18 in the carbonyl position was added to a solution of enzyme, the formation of $H_2{}^{18}O$ was again observed. This demonstrates that the reverse step of Eq. (5-30) occurs, and it implies that the forward step is probably a part of the mechanism for the decarboxylation of the acetoacetate ion. Both labeled compounds slowly form $H_2{}^{18}O$ even in the absence of enzyme because ketones can undergo an uncatalyzed hydration reaction [Eq. (5-32)].

$$R-\overset{\overset{^{18}O}{\|}}{C}-R' + H_2O \rightleftharpoons \underset{1}{R-\overset{\overset{^{18}OH}{|}}{\underset{\underset{OH}{|}}{C}}-R'} \rightleftharpoons R-\underset{\underset{O}{\|}}{C}-R' + H_2{}^{18}O \qquad (5\text{-}32)$$

In intermediate 1, the oxygen-18 atom has a 50:50 chance of leaving as a water molecule.

2. Sodium borohydride is a mild reducing agent which is known to be capable of reducing an imine to give an alkylated amine [Eq. (5-33)].

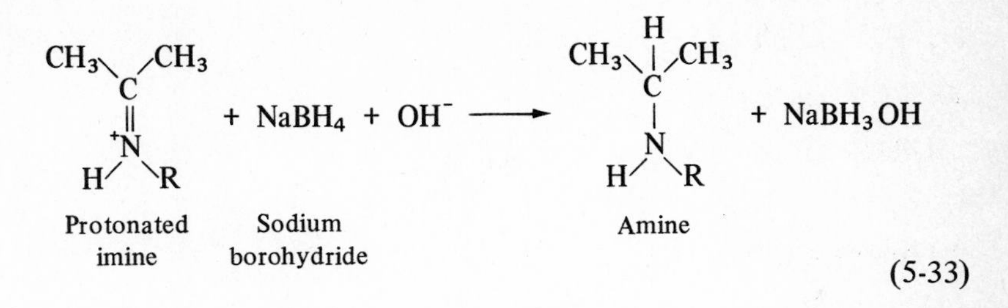

(5-33)

When sodium borohydride was added to a solution containing enzyme and substrate, the enzyme soon became catalytically inactive. Further analysis revealed that one lysine residue in the enzyme had been alkylated by an isopropyl group, as in Eq. (5-33). This demonstrates that the protonated imine produced as a product in Eq. (5-27) is an intermediate in the enzyme-catalyzed reaction, and that a lysine residue in the enzyme provides the catalytic amine group.

3. The reverse steps in Eqs. (5-30) and (5-29) predict that acetone, in which all hydrogen atoms are replaced by deuterium atoms (Fig. 5-3), should lose deuterium ions in the presence of enzyme. Indeed, it has been observed that the enzyme *is* an efficient catalyst for this exchange reaction. This result gives support for Eqs. (5-29) and (5-30) as part of the mechanism of the overall reaction.

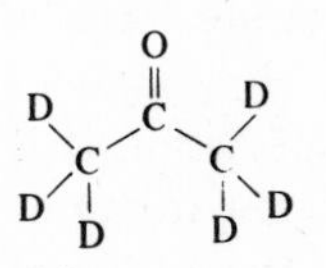

FIGURE 5-3 The structure of deuterated acetone.

Electrophile Catalysis by Pyridoxal Phosphate

The structure of the coenzyme pyridoxal phosphate is shown in Fig. 5-4. The substituents at positions 2 and 5 are believed to be important only for binding the coenzyme to the enzyme. The other groups participate, as we shall see, in the catalytic steps of the reactions

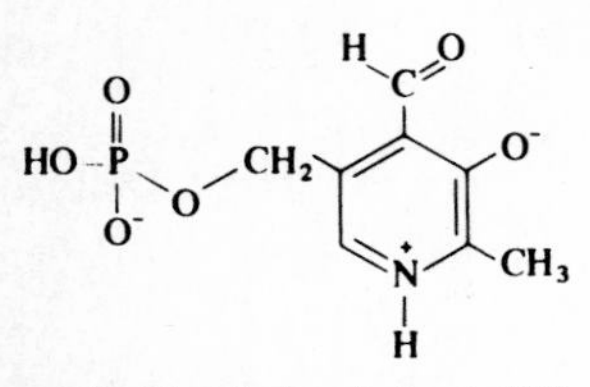

(*a*) Pyridoxal phosphate at pH 7

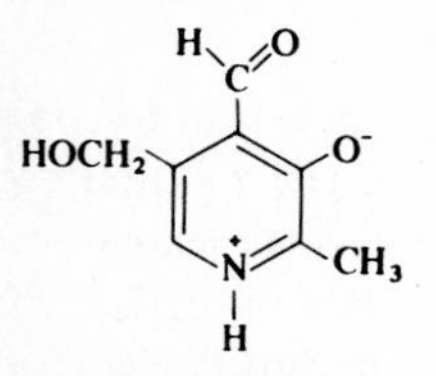

(*b*) Pyridoxal at pH 7

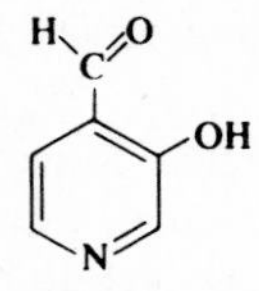

(*c*) 3-Hydroxypyridine-4-aldehyde

FIGURE 5-4 Structures of pyridoxal and related compounds.

catalyzed by pyridoxal phosphate. The coenzyme in its various forms is commonly referred to by nutritionists as vitamin B_6.

Pyridoxal phosphate is a coenzyme of widespread importance. It is required by more than 50 of the known enzymes. Most of the reactions in which it participates involve amino acids as substrates. These reactions may be classified into several types, including transaminations [Eq. (5-34)], racemizations [Eq. (5-35)] (see later), decarboxylations [Eq. (5-36)], and α,β-eliminations [Eq. (5-37)]. All these reactions are important in amino acid metabolism.

Transamination

$$\underset{\text{Oxaloacetate ion}}{{}^{-}\mathrm{O\overset{\overset{O}{\|}}{C}CH_2\overset{\overset{O}{\|}}{C}\overset{\overset{O}{\|}}{C}O^{-}}} + \underset{\text{L-Glutamate ion}}{{}^{-}\mathrm{O\overset{\overset{O}{\|}}{C}CH_2CH_2\overset{\overset{H_3N^{+}}{|}}{C}H\overset{\overset{O}{\|}}{C}O^{-}}} \rightleftharpoons$$

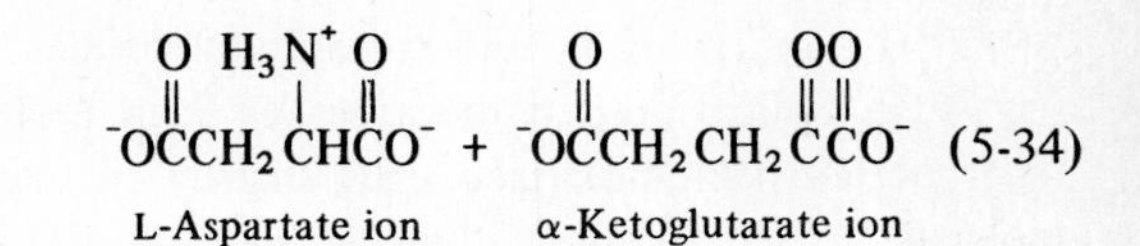

Racemization

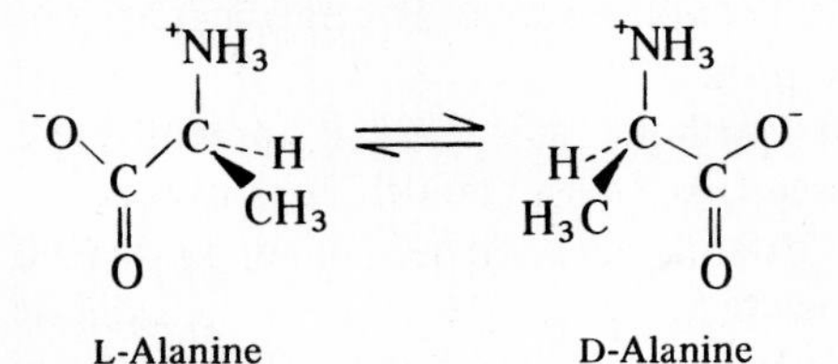

(5-35)

Decarboxylation

$$^{-}OC(=O)CH_2CH(NH_3^{+})C(=O)O^{-} \xrightarrow{+H^{+}} {}^{-}OC(=O)CH_2CH_2NH_3^{+} + CO_2 \tag{5-36}$$

L-Aspartate ion → 3-Aminopropionate ion (β-alanine)

α, β-Elimination

$$HO{-}CH_2CH(NH_3^{+})CO_2^{-} \rightleftharpoons CH_3C(=O)CO_2^{-} + NH_4^{+} \tag{5-37}$$

L-Serine ⇌ Pyruvate ion

The enzyme aspartate aminotransaminase, which catalyzes the reaction in Eq. (5-34), is specific for the conversion of oxaloacetate ion to L-aspartate ion. When the latter is in short supply in the organism, it can be synthesized via the forward reaction of Eq. (5-34). When it is in excess supply, it can be broken down by the reverse reaction. Equation (5-34) is general for most of the amino acids, i.e., other enzymes are present in organisms which can synthesize each of the amino acids from the corresponding α-keto acid, and vice versa. For most of these enzymes the L-glutamate ion is the amino group (nitrogen) donor as in Eq. (5-34).

The conversion of L-alanine to D-alanine [Eq. (5-35)] is the first step in the energy-producing degradation of this particular amino acid.

A 50:50 mixture of L- and D-alanine is called a racemic mixture and shows no optical activity because for each molecule of the L isomer that rotates plane-polarized light slightly in one direction, there is a molecule of the D isomer that has an equal but opposite effect. Since the equilibrium constant for Eq. (5-35) is unity, the enzyme which catalyzes this reaction, alanine racemase, produces a racemic mixture. Therefore, the reaction is a racemization reaction. Racemizing enzymes for other amino acids are also known.

Pyridoxal phosphate alone, without any enzyme present, can catalyze the reactions we are discussing. It is a "model" for the enzyme, although much less efficient. Most detailed model studies have used pyridoxal or 3-hydroxypyridine-4-aldehyde (Fig. 5-4) as a catalyst instead of pyridoxal phosphate. In the absence of enzyme all three are approximately equally efficient. These model studies form the basis for the following discussion.

We shall now center our attention on the transamination reaction of Eq. (5-34). Whether in the presence of pyridoxal phosphate plus enzyme or in the presence of pyridoxal alone, this reaction proceeds in two stages or two half-reactions, Eqs. (5-38) and (5-39). Both half-reactions are mechanistically similar.

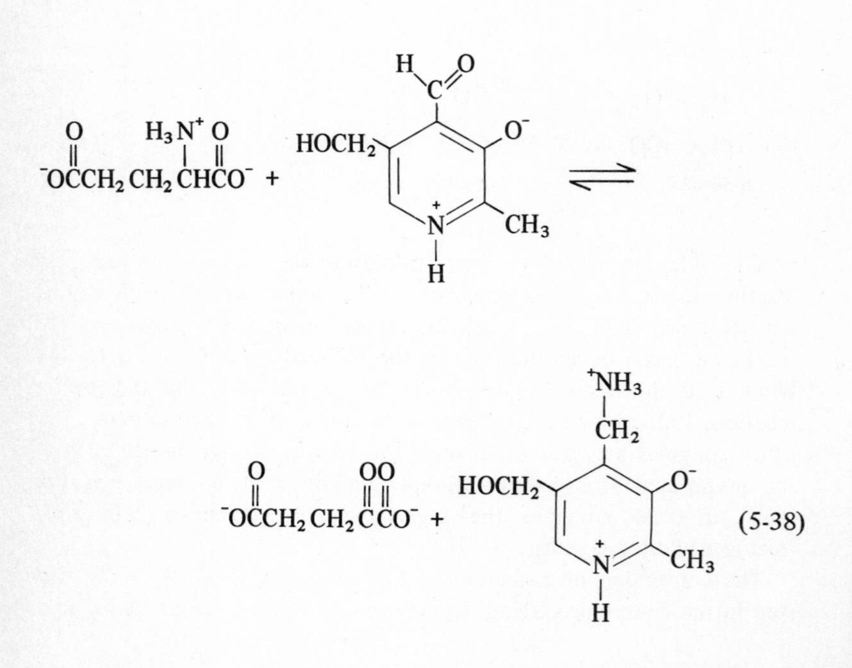

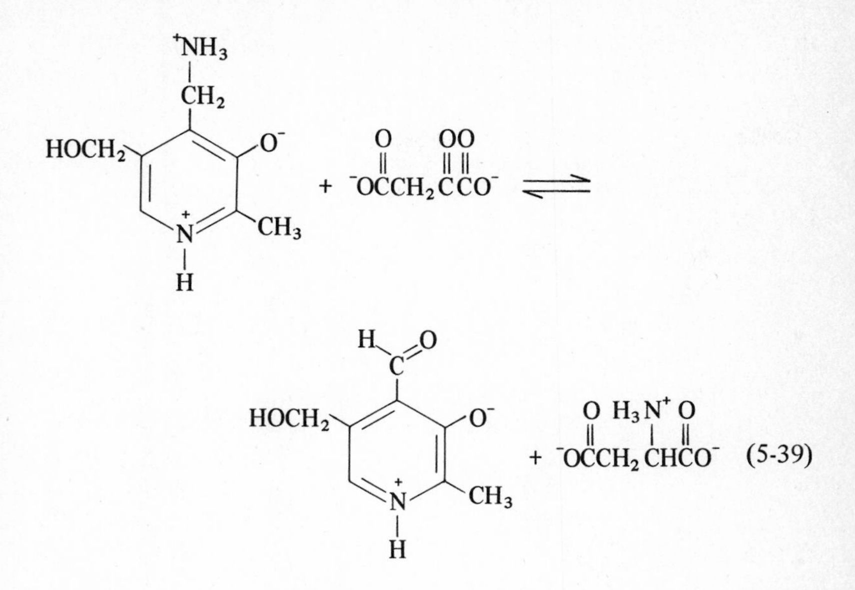

In Fig. 5-5 the intermediate steps are shown for the first half-reaction. It consists of three stages–imine formation to an aldimine, tautomerization to a ketimine, and imine hydrolysis. Since imine hydrolysis is just the reverse of imine formation, the total reaction is highly symmetric. An aldimine is an imine formed between an aldehyde and an amine; a ketimine is an imine formed from a ketone and an amine.

In our discussion of imines in the beginning of this section we noted that imine formation may be catalyzed by general acids and general bases. Pyridoxal provides its own general base catalyst, the ionized hydroxyl group which assists the removal of a proton from the nitrogen atom during dehydration of the carbinolamine species 1, as shown in Fig. 5-5. The aldimine product is stabilized by the hydrogen bond between the hydroxyl group and the imine nitrogen atom.

The conversion of the aldimine in Fig. 5-5 to a ketimine requires the removal of a proton from one carbon atom and the addition of a proton to another carbon atom with the simultaneous migration of a double bond, a process we called tautomerization earlier. It is the rate-determining step in the overall reaction in Fig. 5-5; unlike proton transfers to and from oxygen and nitrogen atoms, proton transfers to and from carbon atoms are slow processes. The tautomerization step is susceptible to catalysis by a general base and by the internal hydroxyl group acting as a general acid. This is shown explicitly in Eq. (5-40),

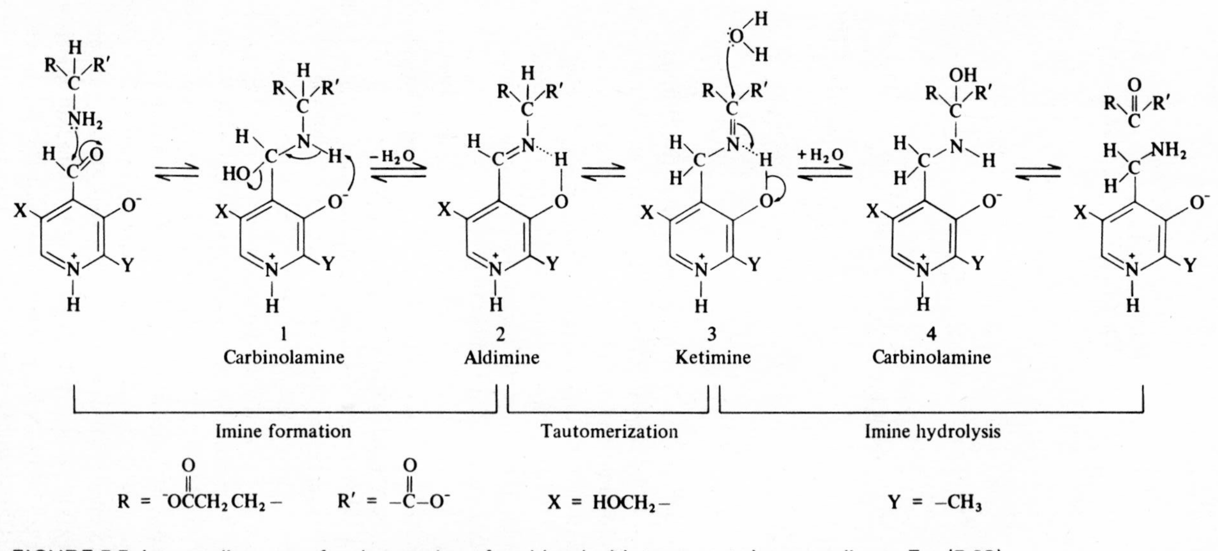

$R = {}^{-}O\overset{O}{\overset{\|}{C}}CH_2CH_2-$ $\quad R' = -\overset{O}{\overset{\|}{C}}-O^-$ $\quad X = HOCH_2-$ $\quad Y = -CH_3$

FIGURE 5-5 Intermediate steps for the reaction of pyridoxal with L-aspartate ion according to Eq. (5-38).

where the second step is a very fast proton transfer to give back the hydroxyl group which is hydrogen-bonded to the imine nitrogen atom.

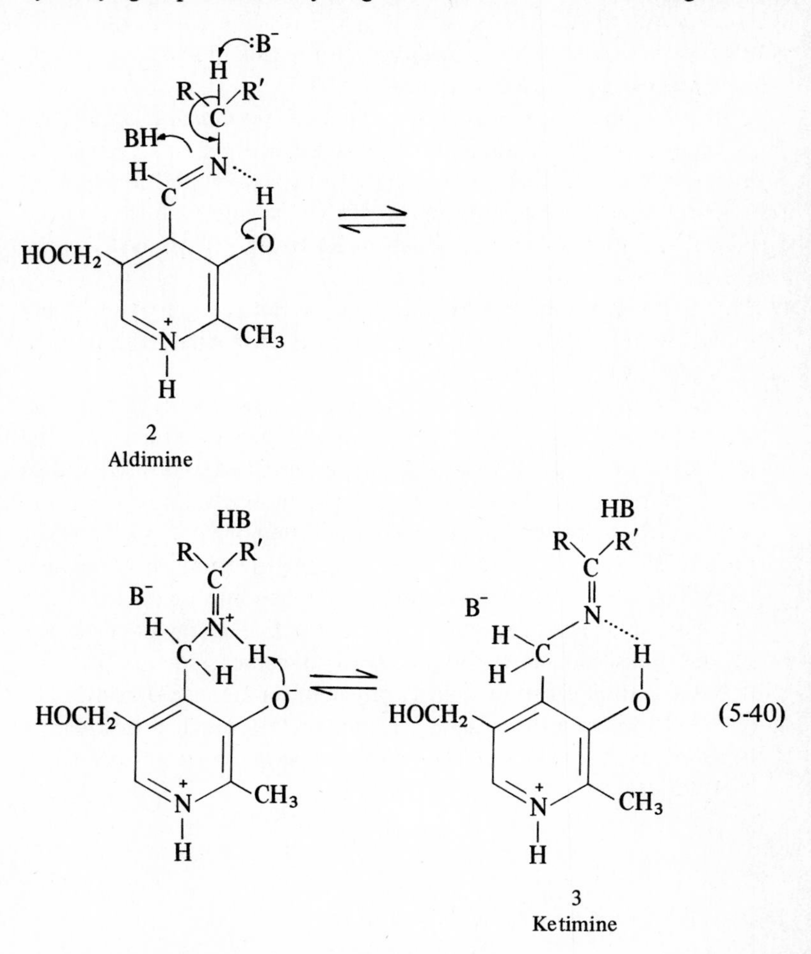

Finally, the imine hydrolysis stage in Fig. 5-5 is mechanistically the reverse of imine formation except that a different imine bond is broken. The hydroxyl group acts as a general acid, and, as we noted earlier in this section, the reaction is also subject to catalysis by added general bases.

In summary, we can say that the reaction of L-glutamate ion with pyridoxal proceeds via the various steps shown in Fig. 5-5. All of these steps may be catalyzed by general acids and/or bases. In this reaction pyridoxal is a reactant, not a catalyst. However, the pyridoxamine produced in the half-reaction of Eq. (5-38) can react according to Eq. (5-39) to give back pyridoxal. The detailed mechanism of this latter reaction is exactly the reverse of Fig. 5-5. Thus Eqs. (5-38) and (5-39) added together constitute a transamination reaction in which pyridoxal, with or without an enzyme, is a catalyst.

In the first half-reaction of transamination, pyridoxal is an electrophile catalyst, i.e., its carbonyl carbon atom accepts an electron pair from the amino group of the substrate in the initial addition step. In the second half-reaction, pyridoxamine is a nucleophile catalyst. Pyridoxal has several features which make it a good catalyst for the transamination reaction.

1. The hydroxyl group is ideally located to provide general acid and general base catalysis. Being intramolecular, such catalysis is particularly effective (Sec. 3-10).
2. The positively charged nitrogen atom in the pyridine ring acts as an electron sink in two ways: (*a*) it tends to pull electrons from the carbonyl carbon atom, thus giving the latter a partial positive charge and making it highly electrophilic; (*b*) in the various transition states, such as that in the tautomerization step of Eq. (5-40) where a carbon-nitrogen bond breaks, the bonding electrons can momentarily roam into the pyridine ring rather than pile up on the single carbon atom. Such a possibility lowers the free energy of activation of the transition state. In fact there is some evidence that the tautomerization reaction actually proceeds via the mechanism of Eq. (5-41), in which species *2a* is an intermediate which forms readily because of the electron-accepting properties of the positively charged nitrogen atom.

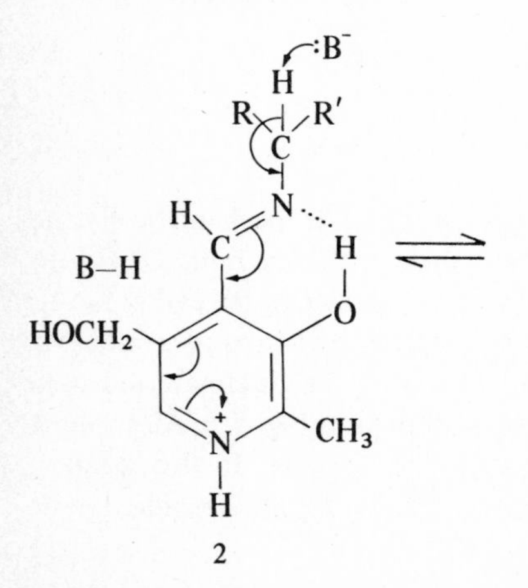

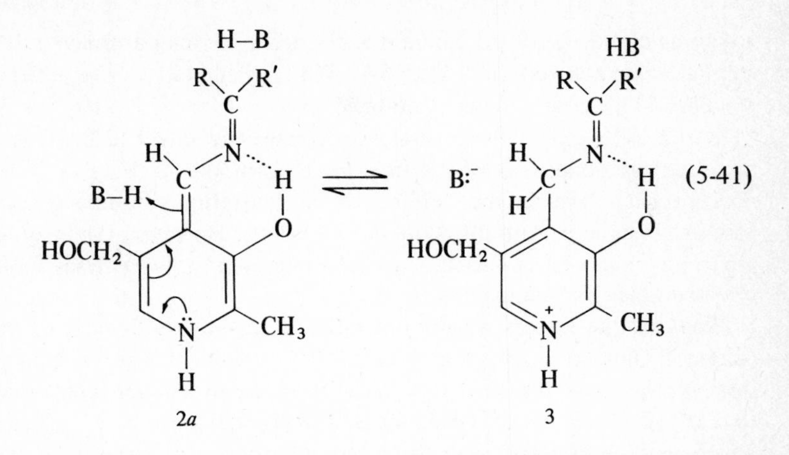

In the enzyme-catalyzed reaction as well as in the absence of enzyme, the tautomerization step [Eq. (5-40) or (5-41)] is the rate-determining step in the transamination half-reaction between an

TABLE 5-1 Some tautomerization rate constants for transamination reactions in the presence of various catalysts

Reaction	Catalyst	Rate Constant, s^{-1}	Rate Constant Relative to Catalysis by Pyridoxal Alone
Tautomerization step in the conversion of:			
(*a*) L-alanine → pyruvate ion (pH 7)	-Pyridoxal alone	1×10^{-5}	1
	-Pyridoxal plus 10 *M* imidazole (in basic form)	1×10^{-4}	10
(*b*) L-glutamate ion → α-ketoglutarate ion (pH 6)	-3-Hydroxypyridine-4-aldehyde only	1×10^{-5}	1
	-3-Hydroxypyridine-4-aldehyde plus 10 *M* imidazole (in basic form)	1.2×10^{-3}	120
	-Pyridoxal phosphate plus the enzyme, aspartate amino-transaminase	3000	300,000,000
	-Pyridoxal plus 0.01 *M* Al^{3+}	1×10^{-4}	10

amino acid and pyridoxal. Some rate constants for the tautomerization step are summarized in Table 5-1. With pyridoxal (or 3-hydroxy-pyridine-4-aldehyde) alone, the rate constant is $1 \times 10^{-5}\ s^{-1}$. Imidazole as a general base catalyst increases the rate 120 times at a hypothetical concentration of the basic form of 10 *M*. The latter concentration level is the "effective" concentration of some general base catalysts in certain intracomplex reactions. However, the enzyme aspartate aminotransaminase is greater than one million times more effective than 10 *M* imidazole.

What are the factors which contribute to this high efficiency of the enzyme? Quite probably, the general base at the active site is held in just the right position so that it has a considerably greater effect than that of 10 *M* imidazole. Also, the active site may have a general acid which is more effective than the hydroxyl group of pyridoxal. In this regard, metal ions such as Al^{3+} have a considerable catalytic effect (see Table 5-1) because they can complex to the imine, as shown in Eq. (5-42). Not only does the metal ion act as a general acid catalyst (or super acid catalyst—see Sec. 4-2), it also holds the whole conjugated system planar, thus facilitating the electron rearrangement which occurs during tautomerization, as shown in Eq. (5-42).

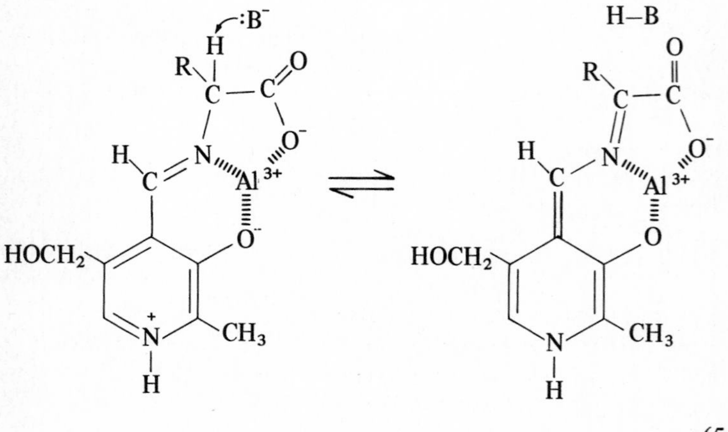

(5-42)

Although aspartate aminotransaminase does not contain a metal ion, the enzyme may play the same role as the metal ion, namely, provide a good general acid catalyst and hold the imine substrate in a planar conformation.

We should also point out that, in the enzyme-catalyzed reactions, pyridoxal phosphate is actually covalently bonded to the enzyme. It forms an imine with the ϵ-amine group of a lysine residue in the enzyme. Thus imine formation between substrate and enzyme-bound pyridoxal is really a transimination reaction and occurs as shown in Eq. (5-43), not as in Fig. 5-5.

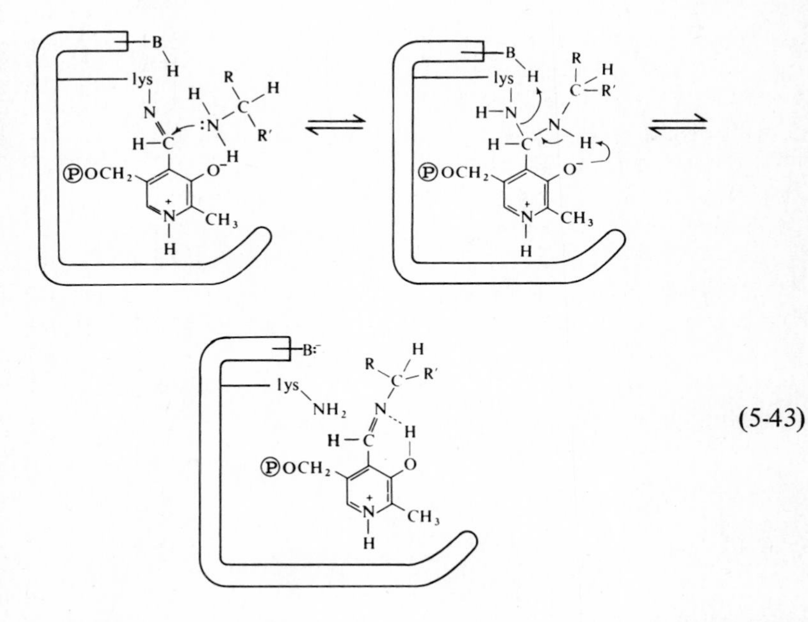

(5-43)

Thus far we have dealt exclusively with the transamination reaction. The other classes of reactions catalyzed by pyridoxal phosphate-requiring enzymes probably follow very similar mechanisms. Figure 5-6 illustrates how each reaction class is believed to occur using the amino acid L-serine for illustrative purposes. Enzymes are known which will catalyze all six classes of reactions shown in Fig. 5-6, if not with L-serine as substrate, then with one or another of the other amino acids. Figure 5-6 is simplified in that imine formation and imine hydrolysis at the beginning and end of each reaction sequence are shown as single steps. All reaction sequences begin at the middle left with the formation of the aldimine between pyridoxal-enzyme and the amino acid substrate. All reaction sequences terminate with the reappearance of pyridoxal phosphate-enzyme (except for the transamination half-reaction which gives pyridoxamine phosphate-enzyme) which to

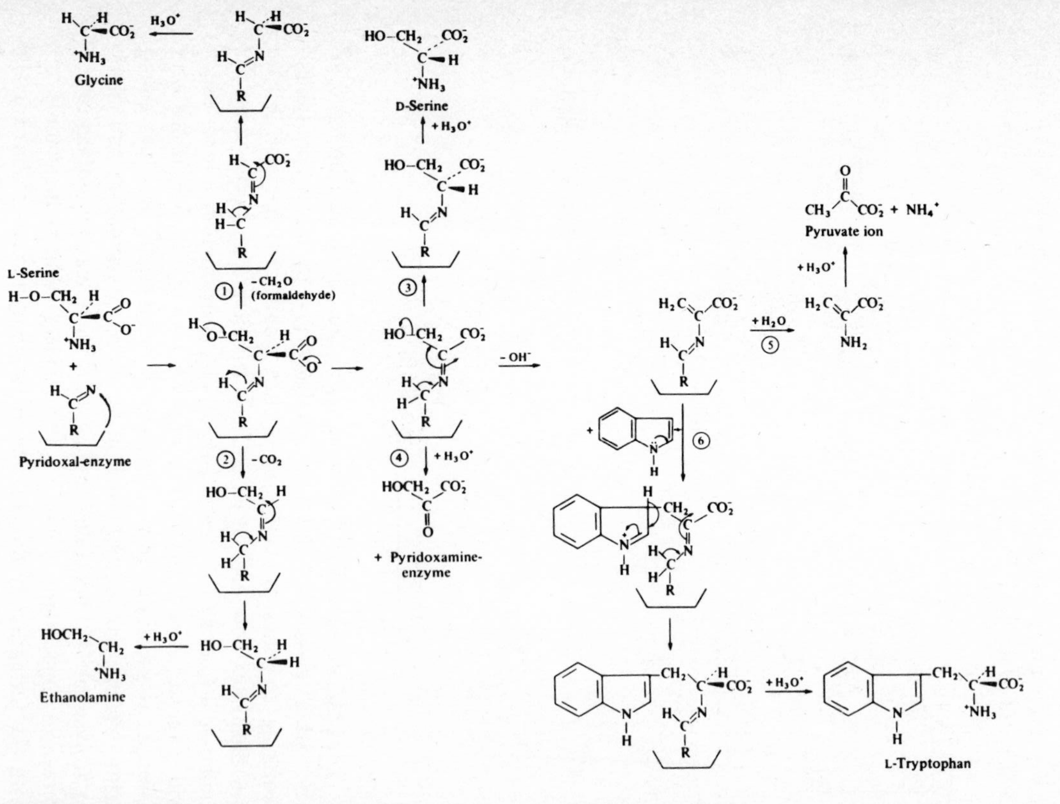

FIGURE 5-6 Possible reactions of amino acids with pyridoxal phosphate and corresponding enzymes. (The enzyme surface is symbolized by the solid line.) (1) α,β-Cleavage; (2) decarboxylation; (3) racemization; (4) transamination half-reaction; (5) α,β-elimination; (6) β-substitution.

save space has not been shown. All arrows point only in the direction which gives rise to the desired product. However, it should be realized that almost all of the steps are reversible. The intermediate above 2 can react in more than one way, leading to decarboxylation, racemization, or elimination, which is pertinent to tryptophan synthesis. To simplify the diagram, protons are made to appear and disappear as needed with no indication of their origin or fate which are either water molecules or acid and base groups on the enzyme. Also, in reaction 6, indole must be present in the solution to give the new amino acid tryptophan. Although mechanisms 1, 2, 3, 5, and 6 in Fig. 5-6 have not been as firmly established as the transamination half-reaction, their similarity to the latter and other known reactions makes the organic chemist feel that they are probably correct.

If one took any amino acid and added it to a solution of pyridoxal, presumably all of the applicable reactions of Fig. 5-6 (for alanine, reactions 2, 3, and 4) would occur to a greater or lesser extent. With only a few exceptions each of the pyridoxal phosphate-enzymes is specific for a given amino acid and a given reaction, such as racemization or transamination but not both; i.e., the enzyme binds the substrate in such a way that one of its reactions is catalyzed to a much greater degree than any of the other possible reactions. Thus the enzyme is not only a catalyst; it also directs or controls the reaction.

SUGGESTED READINGS

Bender, M. L.: "Mechanisms of Homogeneous Catalysis from Protons to Proteins," John Wiley & Sons, Inc., New York, 1971.

Mahler, H. R., and E. H. Cordes: "Biological Chemistry," 2d ed., Harper and Row, New York, 1971.

Lehninger, A. L.: "Biochemistry," Worth Publishers, Inc., New York, 1970.

Ingold, C. K.: "Structure and Mechanism in Organic Chemistry," 2d ed., Cornell University Press, Ithaca, N.Y., 1969.

Breslow, R.: The Mechanism of Thiamine Action, *Journal of the American Chemical Society*, vol. 80, p. 3719, 1958.

Metzler, D. E., M. Ikawa, and E. E. Snell: A General Mechanism for Vitamin B_6-Catalyzed Reactions, *Journal of the American Chemical Society*, vol. 76, p. 648, 1954.

Jencks, W. P.: Mechanism and Catalysis of Simple Carbonyl Group Reactions, *Progress in Physical Organic Chemistry*, vol. 2, p. 63, 1964.

SIX
THE BASIS OF ENZYME ACTION

6-1 SPECIFICITY OF ENZYME-CATALYZED REACTIONS

In previous chapters we have noted that enzymes are good catalysts because they have several catalytic groups acting in concert at the active site. Another important aspect of an enzyme is its specificity, the ability to catalyze the reaction of certain substrates (called good, or specific, substrates) and not other, similar, compounds (called poor or nonsubstrates). In this section we shall analyze two aspects which are doubtlessly involved in determining enzyme specificity: orientation and strain.

Orientation in a Model, Nonenzymic Case

It is intuitively obvious that a catalytic group in a molecule is of no use if it is not correctly oriented to attack the reactive part of the molecule. The importance of proper orientation is vividly illustrated by the relative reactivity of the compounds listed in Table 6-1. They are all dibasic carboxylic acids in which one of the carboxyl groups has been esterified by a substituted phenyl group. The reaction in question is the intramolecular attack by the free carboxylate anion on the ester function to form the corresponding anhydride and release the substituted phenol.

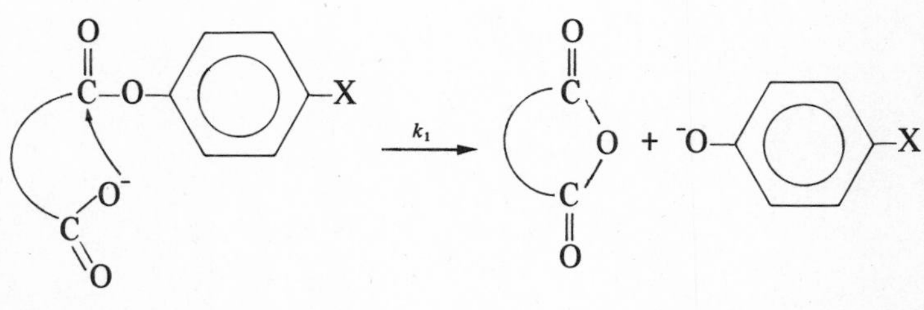

(6-1)

Since the carboxylate ion is later regenerated when the anhydride hydrolyzes, it is properly called a catalytic group. The series of compounds in Table 6-1 may be viewed as a *model* for the importance of proper orientation of substrate and catalytic groups at the active site of an enzyme.

The likelihood of Eq. (6-1) occurring is determined by, among other things, the probability that the carboxylate ion lies in the proper relative position so that it can attack the carbonyl carbon of the ester group. In compound 1 (Table 6-1), derived from glutaric acid, the probability is small because of the freedom of rotation about the four single carbon-carbon bonds connecting the catalyst and the reacting group. In compound 2 the probability of correct orientation is increased, there being one less C–C bond about which the reactive groups can rotate away from each other. In compound 3, rotation cannot occur about the double bond; thus the reactive groups are constrained to lie in close proximity, although even in this case

TABLE 6-1 Relative rates of anhydride formation by monoesters of some dicarboxylic acids

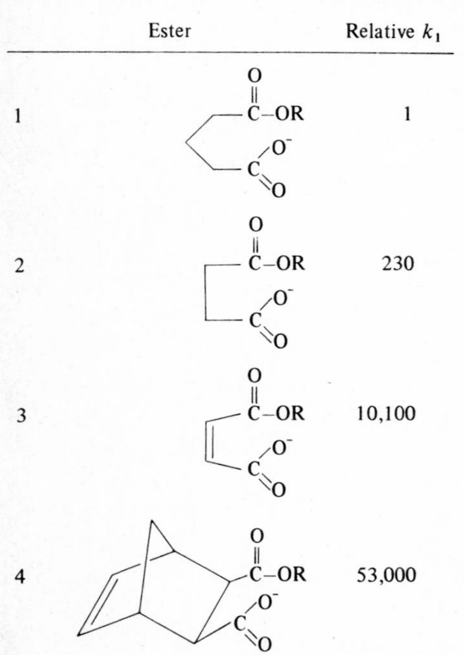

	Ester	Relative k_1
1		1
2		230
3		10,100
4		53,000

orientation is far from ideal, since rotation about two single carbon-carbon bonds can still occur. Finally a molecular model of compound 4 shows that it holds the reactive groups in a slightly more favorable orientation than compound 3, and this is reflected in a slightly faster reaction.

The actual geometry for reaction in Eq. (6-1) is for the nucleophilic oxygen atom of the carboxylate anion to approach the carbonyl carbon of the ester group from a direction perpendicular to the plane of the latter. We could anticipate that the rate enhancement would be even more striking if the two reactive groups could be immobilized in this perpendicular orientation. Even compound 4 does not have such a constraint. This, presumably, is the type of additional influence which an enzyme can provide.

We can summarize the results of Table 6-1 by noting that the reaction rate increases as the probability increases that the reactant and catalytic group are held in the correct conformation or orientation for reaction to occur. In thermodynamic terms, the entropy of the initial states is successively decreased and, therefore, the decrease in entropy on going to the transition state is less for each successive compound in Table 6-1. That is, $\Delta S^\ddagger$ becomes more and more positive; therefore, $\Delta G^\ddagger$ becomes smaller and smaller.

Orientation in Chymotrypsin-catalyzed Reactions

Transferring this idea of the importance of orientation to enzyme-catalyzed reactions, we recognize that a reaction will be most effectively catalyzed if the substrate fits into, and is firmly held at, the active site of an enzyme in such a way that its reacting group is ideally oriented with respect to the catalytic groups of the enzyme.

There are two aspects to this idea of substrate binding. First, a good substrate must be held in an orientation in which its reacting group is properly aligned with the catalytic groups. Second, the larger the fraction of time the substrate is so held at the active site, the more likely it is to react and the better a substrate it is. We shall refer to the latter aspect as tightness of binding. A tightly bound substrate is one which has a small dissociation constant, K_s.

These two aspects of specificity, tightness of binding and orientation, are usually referred to as *binding specificity* and *kinetic specificity*, respectively. The compound *N*-acetyl-L-tryptophan ethyl ester is a good substrate of chymotrypsin with a K_s of 2.5×10^{-3} M and a k_{cat} of 27 s^{-1} at 25°C. These values reflect good binding specificity and good kinetic specificity. On the other hand, the mirror image *N*-acetyl-D-tryptophan ethyl ester hardly reacts at all in the presence of chymotrypsin. Nevertheless, it binds to the active site with a dissociation constant of 0.8×10^{-3} *M*, even slightly better than the L isomer. Therefore, it has good binding specificity, but extremely poor kinetic specificity. Presumably, the D isomer is tightly held, but in the wrong orientation for reaction to occur.

We shall now look at specificity in chymotrypsin-catalyzed reactions in more detail. The general formula for simple substrates is given in Eq. (6-2), where X stands for the amide or ester group, R_1 for an acyl

TABLE 6-2 Overall reactivity for the chymotrypsin-catalyzed hydrolysis of methyl esters of some *N*-acetyl-L-amino acids

Acyl group	Side-chain Structure (R_2)	$\frac{k_{cat}}{K_m}$ $M^{-1} \cdot s^{-1}$
N-Acetyl-L-tryptophanyl-	(indol-3-yl)$-CH_2-$	420,000
N-acetyl-L-tyrosyl-	$HO-C_6H_4-CH_2-$	365,000
N-acetyl-L-hexahydrophenylalanyl-	$C_6H_{11}-CH_2-$	80,000
N-acetyl-L-phenylalanyl-	$C_6H_5-CH_2-$	42,000
N-acetyl-L-methionyl-	$CH_3-S-CH_2CH_2-$	2,300
N-acetyl-L-leucyl-	$(CH_3)_2CH-CH_2-$	1,590
N-acetyl-*S*-methyl-L-methionyl-	$(CH_3)_2\overset{+}{S}-CH_2CH_2-$	0.21
N-acetyl-L-glutamyl-	$^-O_2C-CH_2CH_2-$	0.035
N-acetyl-glycyl-	H–	0.0098

substituent on the α-amino group, and R_2 for the side chain which determines the identity of the amino acid.

$$\begin{array}{l} \quad\ \ /H \\ R_1-N \quad\ O \\ \quad\ \ \diagdown \quad \| \\ \quad H-C-C-X \\ \qquad\ \ | \\ \qquad R_2 \end{array} \tag{6-2}$$

In Table 6-2 are listed data for the overall rate of reaction (k_{cat}/K_m) for several substrates of chymotrypsin in which only R_2 is varied. [See Sec. 2-4, particularly Eq. (2-21), for the meaning of k_{cat} and K_m.] Two points are clear from these data. (1) The side chain R_2 must be nonpolar for the substrate to be a good one. When a positive or negative charge is present, the reaction is very slow. (2) The larger (or bulkier) the nonpolar side chain, the better the substrate. This suggests that an apolar interaction exists between R_2 and a region on the active site. The x-ray structure confirms this, revealing a nonpolar crevice on the enzyme surface into which the R_2 group fits.

Data similar to those in Table 6-2 in which R_1 and X are systematically varied show that the nature of these groups has relatively little effect on the reactivity of the substrate.

Thus far we have discussed the overall reactivity of chymotrypsin substrates. However, the overall reaction involves several steps. To understand what contributes to specificity it is better to look at a single reaction step. Consequently, we now turn our attention to the deacylation step. [The deacylation step is the reaction of the acyl-enzyme with H_2O—see Eq. (3-51).]

The pertinent data for the deacylation of several α-acetamido-acyl-chymotrypsins are presented in Table 6-3. The column headed "Relative k_3" presents the deacylation rate constants relative to *N*-acetylglycyl-chymotrypsin. The following generalizations can be made concerning the data in Table 6-3. (1) The bulkier the R_2 group, the faster the L isomer deacylates. (2) The bulkier the R_2 group, the slower the D isomer deacylates. (3) The bulkier the R_2 group, the smaller (better) the dissociation constant, K_i, of the corresponding D-amino acid amide. (4) All L isomers deacylate more rapidly than all D isomers, and the *N*-acetylglycyl- and acetyl-chymotrypsins, which lack bulky side chains, react faster than all D isomers and slower than all L isomers.

TABLE 6-3 Relative kinetic parameters for some L- and D-acyl-chymotrypsins and K_i values for the corresponding D-amides

Acyl Group	K_i for D-Amide, M	Relative k_3	$\Delta H^{\ddagger}$, kcal mol^{-1}	$\Delta S^{\ddagger}$, cal $deg^{-1} \cdot mol^{-1}$
N-acetyl-L-tryptophanyl-	......	560.	11.9	−11
N-acetyl-L-phenylalanyl-	......	296.		
N-acetyl-L-leucyl-	......	24.1		
N-acetylglycyl-	0.70	1.0		
N-acetyl-D-leucyl-	0.140	0.166		
N-acetyl-D-phenylalanyl-	0.011	0.047		
N-acetyl-D-tryptophanyl-	0.0015	0.030		
Acetyl-	......	0.185	11.1	−31

These observations may be explained in terms of the following model of three-point acyl interactions at the chymotrypsin active site. *Interaction A*: All acyl groups are covalently attached by an ester bond to serine-195 of chymotrypsin. Thus there is no possibility of complete dissociation as in the enzyme-substrate complex until the deacylation reaction occurs. *Interaction B*: The amino acid side chain R_2 is bound via an apolar interaction to the active site crevice. *Interaction C*: The α-amino hydrogen atom can hydrogen-bond to some acceptor atom at the active site. (This conclusion is inferred from the markedly lower reactivity of acyl-enzymes in which this hydrogen atom is not present. It also appears in the x-ray structure of crystalline ES complexes.) These interactions are illustrated schematically in Eq. (6-3).

Interaction *A* exists as long as the acyl-enzyme has not deacylated. Interactions *B* and *C*, being noncovalent, will exist intermittently depending upon the strength of their binding attractions. When either or both of the latter two groups dissociate away from the enzyme surface, the orientation of the carbonyl group will change with respect to the catalytic groups. Equation (6-3) also illustrates this aspect. The stronger the *B* and *C* interactions are, the greater the fraction of time the acyl group is held in its particular orientation [form I of Eq. (6-3)], and the faster the deacylation reaction, if that orientation is a good one. Interaction *C* is the same for all acyl groups in Table 6-3 (except that it does not exist for the last compound); only interaction *B* varies.

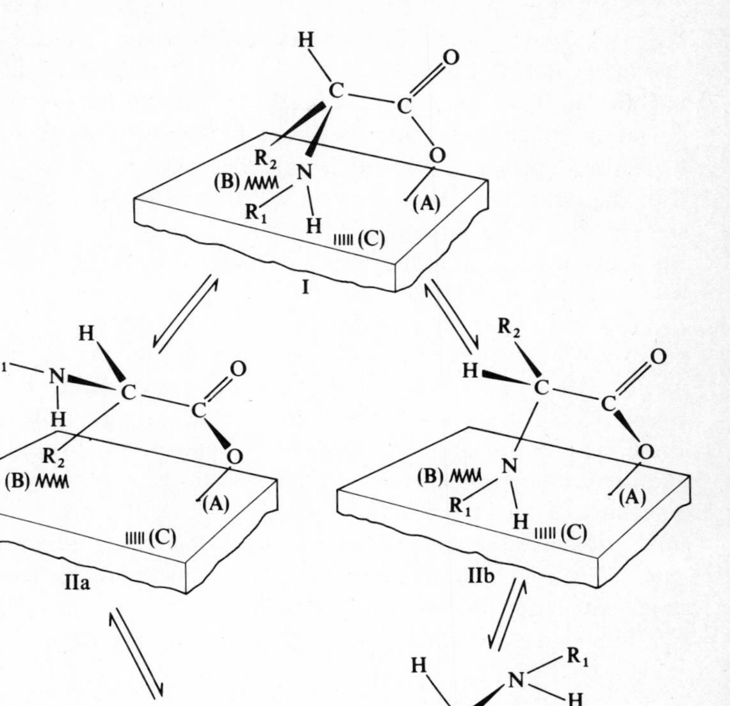

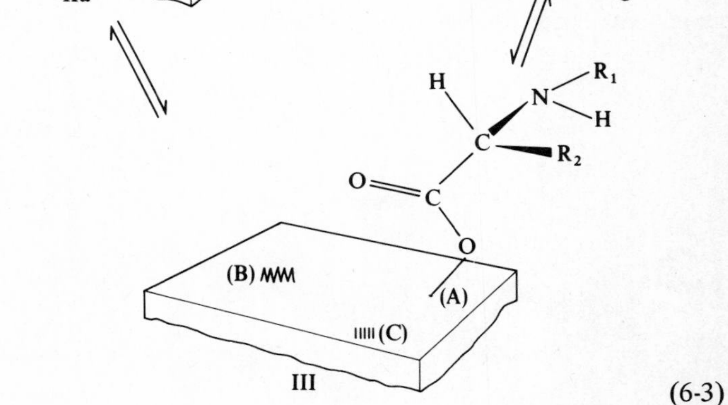

(6-3)

The relative strengths of interaction B for both D- and L-acyl groups is indicated by the relative values of K_i for the corresponding D-amides since D and L isomers generally bind equally strongly. Thus the tryptophan side chain has the strongest interaction (smallest dissociation constant) and the glycine side chain the weakest interaction (largest dissociation constant). In fact, since we have earlier concluded that interaction B is apolar and depends on the bulk of the side chain, the glycine side chain has essentially no interaction of this type. Therefore, for the first four acyl groups in Table 6-3, the relative rate is successively smaller because interaction B becomes successively weaker, with the result that the acyl group is held less, on the average, in the correct orientation for reaction. We are assuming, in saying this, that when the acyl group is fully bound at sites B and C, the substrate is held in an ideal orientation.

Figure 6-1*a* shows one conformation of the *N*-acetyl-L-phenylalanyl group which we shall arbitrarily call ideal; i.e., the carbonyl group is ideally oriented with respect to the catalytic groups (not shown). In the D-isomer, Fig. 6-1*b*, the configuration of the acyl group is changed. From the α-carbon atom of Fig. 6-1, the bonds to the hydrogen atom and the carbonyl carbon atom are reversed in the D isomer (Fig. 6-1*b*) as compared to the L isomer (Fig. 6-1*a*). If, as we assume, the three interactions are the same in both isomers (which is approximately the case in Fig. 6-1), it is clear that the carbonyl group cannot take up the same orientation in the tightly bound D isomer that it does in the tightly bound L isomer. If tight binding of the L-acyl group orients its

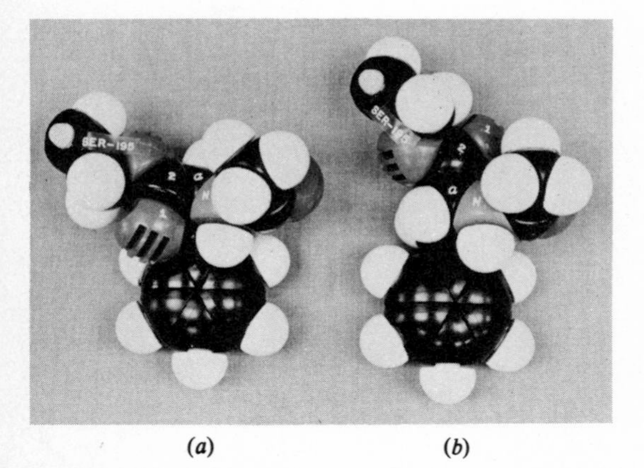

(*a*) (*b*)

FIGURE 6-1 Models of the *N*-acetylphenylalanyl group bound to the active site of chymotrypsin: (*a*) L isomer; (*b*) D isomer. The labeled atom(s) are: Ser-195, the side chain of serine-195 in chymotrypsin; 1, 2, α, N refer to the carbonyl oxygen, the carbonyl carbon, the α-carbon, and the nitrogen atoms, respectively, of phenylalanine.

carbonyl group ideally, tight binding of the D isomer will misalign its carbonyl group with respect to the catalytic groups. Thus the bulkier and tighter-binding R_2 groups will hold the D-acyl group more of the time in an unreactive alignment and the reaction rate will be slower, precisely what we observe.

Figure 6-1 has shown us how specificity leads to stereospecificity. We can summarize with a simple analogy. Imagine your head as the enzyme. The phenyl ring slips into your mouth (crevice) (interaction *B*), your nose makes a hydrogen bond with the amino hydrogen atom (interaction *C*), and the serine-195 side chain projects at some distance from your left temple (interaction *A*). This places the carbonyl oxygen atom (atom 1) of the L isomer below your left eye, where a general acid catalytic group might be aligned with it. The attacking water nucleophile (not shown) would approach towards the bridge of your nose where a projecting general base group could catalyze its attack on the carbonyl carbon atom (atom 2). With the D isomer (Fig. 6-1*b*) it is clear that, given the same interactions as for the L isomer, the carbonyl group would not be properly aligned with the general acid and general base catalytic groups. Proper alignment could occur only if one or both of interactions *B* and *C* were broken so the carbonyl group could rotate into the proper alignment. Thus the tighter the binding interaction, the less likely the D isomer is to assume a reactive orientation.

Finally, the *N*-acetylglycyl and the acetyl entries in Table 6-3 fit into this explanation. The former does not have a significant *B*-type interaction; therefore, it experiences neither the positive orienting effect that the L isomers do nor the negative orienting effect that the D isomers do. Consequently, it reacts at a rate slower than any of the L isomers and faster than any of the D isomers. The acetyl group reacts even more slowly than *N*-acetylglycyl because it lacks even the *C*-type interaction to help orient it.

Another kind of experimental data confirms the idea that fast deacylation occurs because the acyl group is tightly held in the correct orientation for reaction. We recall from Chap. 1 that a reaction rate constant is related to the free energy of activation $\Delta G^{\ddagger}$. The latter in turn is a function of $\Delta H^{\ddagger}$, the enthalpy of activation, and $\Delta S^{\ddagger}$, the entropy of activation. It is not fully clear just what physical interpretation should be given to these last two parameters. However, in many cases it seems that $\Delta S^{\ddagger}$ is predominantly a measure of the probability that the reacting atoms are properly aligned for reaction to occur. On the other hand $\Delta H^{\ddagger}$ is a measure of the energy which must be put into the properly aligned ground state to stretch existing bonds to the breaking point while bringing together atoms that will form new bonds, i.e., the energy required to go from the properly aligned ground state to the transition state.

Entropy decreases as order increases. Thus, if the average orientation (or order) of reacting groups in the ground state is quite different from the orientation in the transition state, $\Delta S^{\ddagger}$ will be large and negative. But in a reaction where the average orientation (over a period of time) in the ground state is more like that in the transition state, $\Delta S^{\ddagger}$ will be less negative. On these grounds one would conclude from the $\Delta S^{\ddagger}$ values in Table 6-3 that the *N*-acetyl-L-tryptophanyl group has an average orientation in the ground state which resembles its orientation in the transition state much more than the ground-state orientation of the acetyl group resembles its transition state. Thus this line of argument brings us to the same conclusion we made earlier that the L-tryptophanyl group deacylates rapidly because it is held in the orientation required for reaction. On the other hand the acetyl group is not so held, and only occasionally happens to assume, momentarily, the correct orientation for reaction. Both acyl-enzymes have nearly the same $\Delta H^{\ddagger}$ indicating that once the correct orientation exists in the ground state, the likelihood of a reaction occurring is the same with both.

Strain

Most chemical reactions involve the breaking of a covalent bond. Strain could increase the rate of a reaction as illustrated schematically in Fig. 6-2, where, as the substrate binds to the enzyme, the bond to be broken is stretched. In essence, the ground state of the strained ES complex

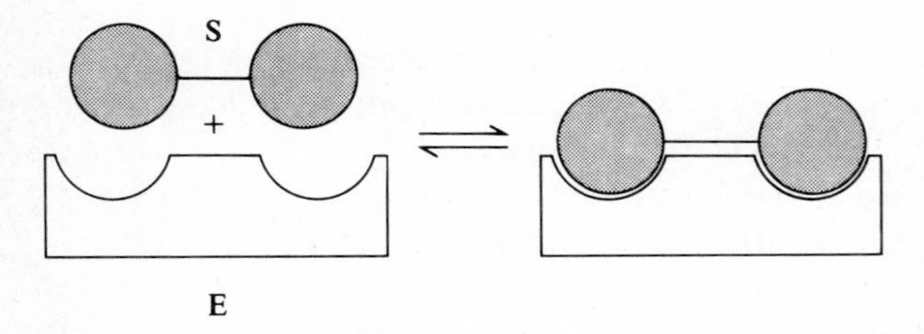

FIGURE 6-2 Schematic representation of how strain may be induced in a substrate when bound to an enzyme. The strained (or stretched) bond is more readily broken than when unstretched.

has become more like the transition state where the bond must be stretched to the breaking point. Strain in this context can mean other things than just the stretching of a bond. It could be the compression of a bond which becomes shorter in the transition state, or it could be the distortion of certain bond angles which makes the molecule more like the transition state.

This idea of strain is difficult to test experimentally, and so it is rather speculative to talk about it. However, there are data which indicate that strain is important in some instances.

$$\mathrm{R{-}\overset{\overset{\displaystyle O}{\|}}{C}{-}OR'} \tag{6-4}$$

One such case is the series of long-chain fatty acid esters as substrates for chymotrypsin. As the length of R [Eq. (6-4)] increases, the overall reactivity of the ester (that is, k_{cat}/K_m) increases. This could mean that k_{cat} increases, or K_m decreases, or both. It would be reasonable to predict that k_{cat} should be the same for all these substrates since the reactive ester group is the same, and that K_m (or K_s) should decrease as was the case in Table 6-3. In fact, however, just the opposite is observed: K_s remains essentially constant and k_{cat} increases as the length of the side chain (R) increases. This is rather perplexing, but the concept of strain can explain it.

To induce strain requires energy. If binding a substrate to the enzyme causes strain to the extent of say 1 kcal mol^{-1}, that energy must come from a correspondingly lesser strength of binding compared to what it would be if no strain occurred.

In the substrate series mentioned, it may be that the expected extra binding energy for the longer chain acids is utilized instead to induce strain in the bound substrate at a strategic point, thus facilitating the reaction and resulting in a larger value of k_{cat} than for the shorter, less strained, bound substrates. Since the potentially available additional binding energy is being used in causing strain, the K_s values remain constant.

One of the appealing aspects of strain is that a small amount of it can have a large effect on a rate constant. Suppose substrate A binds to an enzyme in such a way that strain is induced in A to the extent of 1 kcal mol^{-1}; i.e., if binding could hypothetically occur without strain being induced, the binding interaction would be 1 kcal mol^{-1} stronger than it actually is. If the strain is so induced that it makes the ground state of A become more like the transition state, the free energy of activation, $\Delta G^{\ddagger}$, will have been effectively diminished by 1 kcal mol^{-1}. But the rate constant k is proportional to $e^{-\Delta G^{\ddagger}/RT}$. Therefore, a reduction of $\Delta G^{\ddagger}$ by 1 kcal mol^{-1} corresponds to a rate increase of 5.4 times at 25°C. A reduction of $\Delta G^{\ddagger}$ by 4 kcal mol^{-1} corresponds to a rate increase of $(5.4)^4$, or almost 1,000 times. Of course, the strain induced in a substrate might only partially be applied to lowering $\Delta G^{\ddagger}$, or it might even add to $\Delta G^{\ddagger}$ by forcing A into a less-reactive conformation.

6-2 REGULATION OF ENZYMES

For any complex organism to survive, be it a single-cell amoeba or the human body, the chemical reactions that occur must be under proper control. Just what these control mechanisms are is a subject of much intensive research.

Many of the diseases which we speak of collectively as cancer are failures of control mechanisms, the mechanisms that control the rate of cell division. Our ability to treat such diseases effectively requires that we understand how individual chemical reactions are controlled. Advances will be made in this area by studying first those systems which are relatively simple even though they have no apparent relation to any practical attempt to cure cancer or any other disease.

An example of a very common type of control mechanism is found in the biosynthesis (i.e., synthesis in the living organism) of the amino acids. It is clearly to the organism's advantage if it keeps available a constant supply of each amino acid (for making proteins), making more when the supply is diminished, making less or even stopping production when the supply is adequate or in surplus. One way this goal is achieved is by a process called *feedback inhibition*.

The synthesis of the amino acid L-isoleucine illustrates feedback inhibition. In Fig. 6-3 is the set of reactions leading to the desired product, L-isoleucine. Each reaction is catalyzed by a different enzyme (E_1 through E_5). Although L-threonine is used by the organism in additional ways than just as a precursor of L-isoleucine, the four intermediate compounds leading from L-threonine to L-isoleucine represent a unique pathway whose only purpose is to produce L-isoleucine. Thus, if the cell has an adequate supply of L-isoleucine there is no need to have large amounts of these four compounds around. To produce them would only be a waste of energy. Recognizing this, the organism has

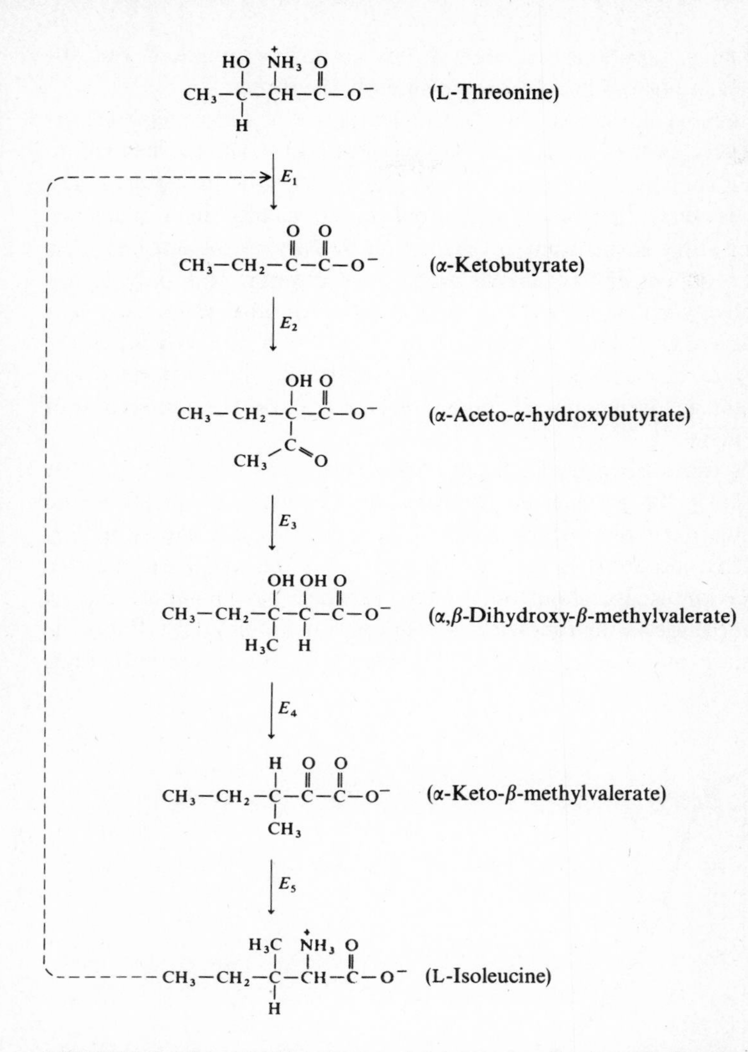

FIGURE 6-3 Biosynthetic pathway of L-isoleucine. E_1 through E_5 are the enzymes catalyzing the different steps. The dashed line represents the effect of L-isoleucine as a feedback inhibitor of E_1, L-threonine deaminase.

developed a feedback control mechanism in which the end product, L-isoleucine, acts as an inhibitor for the first enzyme in the pathway, L-threonine deaminase (E_1). Thus the sequence of reactions is cut off right at the start, and production of L-isoleucine stops. Of course, if the

L-isoleucine supply is depleted, E_1 is no longer inhibited and the reactions in Fig. 6-3 take place, replenishing the supply.

Figure 6-4 illustrates how the concentration of L-isoleucine affects the activity of the enzyme L-threonine deaminase. The curve is said to have sigmoidal shape because of its resemblance to the letter S. The shape is such that only a fivefold or so change in L-isoleucine concentration is sufficient to change the activity of the enzyme from almost 100 percent to almost zero, or vice versa. Not only is this control very efficient; it is also very specific. No other amino acid—not even L-leucine which is similar in structure—has any effect on the activity of this enzyme. Clearly this is desirable; the organism would not want an oversupply of, say, L-leucine to stop the production of L-isoleucine.

How can we explain feedback inhibition on a molecular level? That is, what is the mechanism of interaction between L-isoleucine and threonine deaminase which leads to the activity curve shown in Fig. 6-4? This question has not been answered for this particular enzyme, but our knowledge about other enzymes which have identical control behavior suggests the following mechanism. First of all, the L-threonine deaminase molecule consists of four identical protein subunits. Each

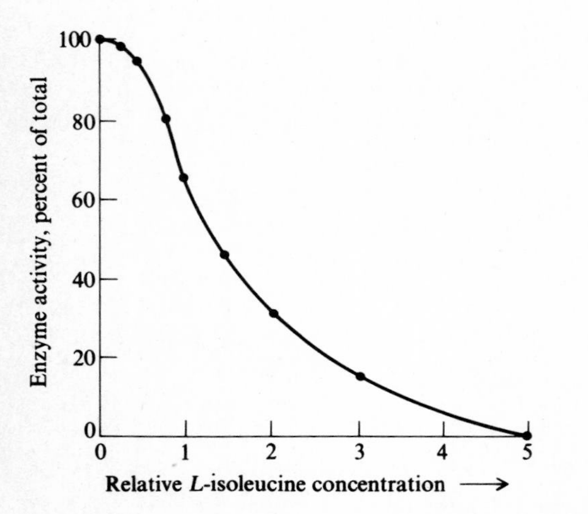

FIGURE 6-4 Effect of L-isoleucine concentration on the catalytic activity of L-threonine deaminase. [*From J. P. Changeux, The Control of Biochemical Reactions, Scientific American, vol. 212, April, p. 40 (1965).*]

subunit has an active site where the substrate L-threonine binds and undergoes the deamination reaction. Each subunit also has a second site, distant from the active site, which is a potential binding site for L-isoleucine. However, L-isoleucine can bind to this site only if the subunit changes to a new three-dimensional conformation, a conformation in which the active site is no longer capable of binding L-threonine substrate molecules. One hypothesis, which seems valid for many enzymes like L-threonine deaminase, holds that all protein subunits in the tetrameric molecule must have the same conformation, and that the collection of enzyme molecules is a mixture of these two conformational forms in equilibrium with each other. These ideas are pictured schematically in Fig. 6-5. The dominant conformational state is, let us say, that capable of binding substrate (form R in the figure), and catalyzing its reaction. However, there will always be some of form T present to which L-isoleucine can bind. If the concentration of L-isoleucine is high enough, it will, by binding to the enzyme, shift the equilibrium in favor of the T state. Because conformation T is catalytically inactive, the deamination of L-threonine stops and L-isoleucine has exerted its feedback inhibitory effect. This type of mechanism, unlike simple competitive inhibition, predicts the sigmoid behavior observed in Fig. 6-4, an inhibition which is much more sensitive to the controlling inhibitor concentration than competitive inhibition. Enzymes which behave this way are often called *allosteric enzymes*.

6-3 MULTIENZYME SYSTEMS

There are many multienzyme systems. They are involved in the biosynthesis of fatty acids, in the biosynthesis of lactose (milk sugar), in the biosynthesis of proteins which involves ribosome bodies in the cell as well as many other entities, in the biosynthesis of the amino acids cysteine and tryptophan, and with many allosteric enzymes, which contain both a catalytic subunit and a regulatory subunit (Sec. 6-2).

In the allosteric enzyme aspartate transcarbamylase, the catalytic and regulatory subunits (which regulate the activity) can be separated from one another and put back together. When they are separated, the catalytic subunit has nonsigmoid kinetic properties; when the regulatory subunit is attached, sigmoid kinetics are observed.

There are many enzymes which are composed of subunits. For

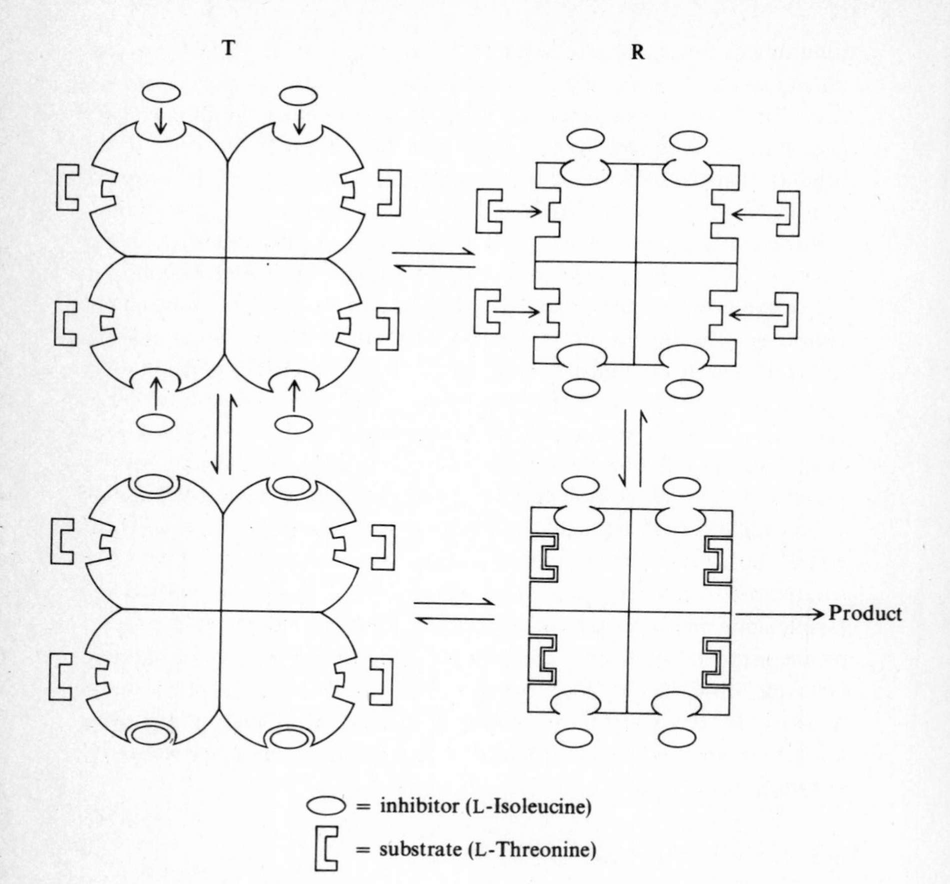

FIGURE 6-5 Schematic representation of the two conformational states of L-threonine deaminase—one which binds inhibitor (T state) and one which binds substrate and catalyzes its reaction (R state).

example, the enzyme muscle aldolase, which catalyzes an aldol condensation in the sugar series, has a molecular weight of 160,000 and consists of four subunits of molecular weight 40,000 each. Also the enzyme yeast alcohol dehydrogenase, which is involved in the oxidation-reduction reaction between ethanol and acetaldehyde, has a molecular weight of 150,000 and consists of four subunits of molecular weight 37,000 each. But in these systems the subunits are presumably identical, whereas in multienzyme systems which are also composed of multiple subunits, the components (subunits) are usually different.

One of the important multienzyme systems is the fatty acid synthetase complex. The biosynthesis of long-chain fatty acids is a multistep process in which eight different catalytic activities participate. In lower species, such as bacteria, the various enzymes can be studied separately, but in higher species the enzymes are contained in macromolecular complexes referred to as fatty acid synthetases. The synthetase from baker's yeast was isolated as discrete particles with a molecular weight of 2.3 million and was recently crystallized, a remarkable feat. Electron micrographs of yeast synthetase show particles of 210 to 250 Å in diameter with a high degree of structural order. The complex can be dissociated into many subunits with an average molecular weight of about 100,000. The complex is made up of eight enzymes in three sets, there being a total of 24 enzymes in all.

On the basis of experiments using radioactive isotopes as tracers, and other evidence, Lynen, a Nobel prize winner, has proposed that the biosynthesis of long-chain fatty acids in yeast occurs by a series of reactions in which the central and peripheral sulfhydryl groups (those which are relatively unreactive with iodoacetamide and those which are readily alkylated by iodoacetamide, respectively) have functional roles as acyl group carriers. This is illustrated in Scheme 1.

The sequence begins with transfer of the acetyl group (a two-carbon acid) from acetyl coenzyme A (a thiol ester) to the central sulfhydryl group, after which it is transferred to the peripheral sulfhydryl group (steps 1 and 2, respectively).

The first step in chain elongation involves a transfer of the malonyl group (a three-carbon diacid) from malonyl coenzyme A (another thiol ester) to the central sulfhydryl group (step 3). This is followed by a condensation reaction in which the protein-bound acetyl and malonyl groups interact to form carbon dioxide and an acetoacetyl protein derivative (step 4). In subsequent steps the β-ketoacyl protein derivative undergoes stepwise reduction (step 5, NADPH is a biological reductant), dehydration (step 6), and further reduction (step 7) to form finally the butyryl-enzyme derivative. The butyryl group is finally transferred to the peripheral sulfhydryl group (step 8, analogous to step 2), and the chain-lengthening sequence begins again with the attachment of another malonyl group to the central sulfhydryl group. The synthesis of fatty acids thus involves a series of reactions in which two-carbon units are added at a time. In many ways the biosynthesis of

SCHEME 1† Fatty acid synthetase of yeast

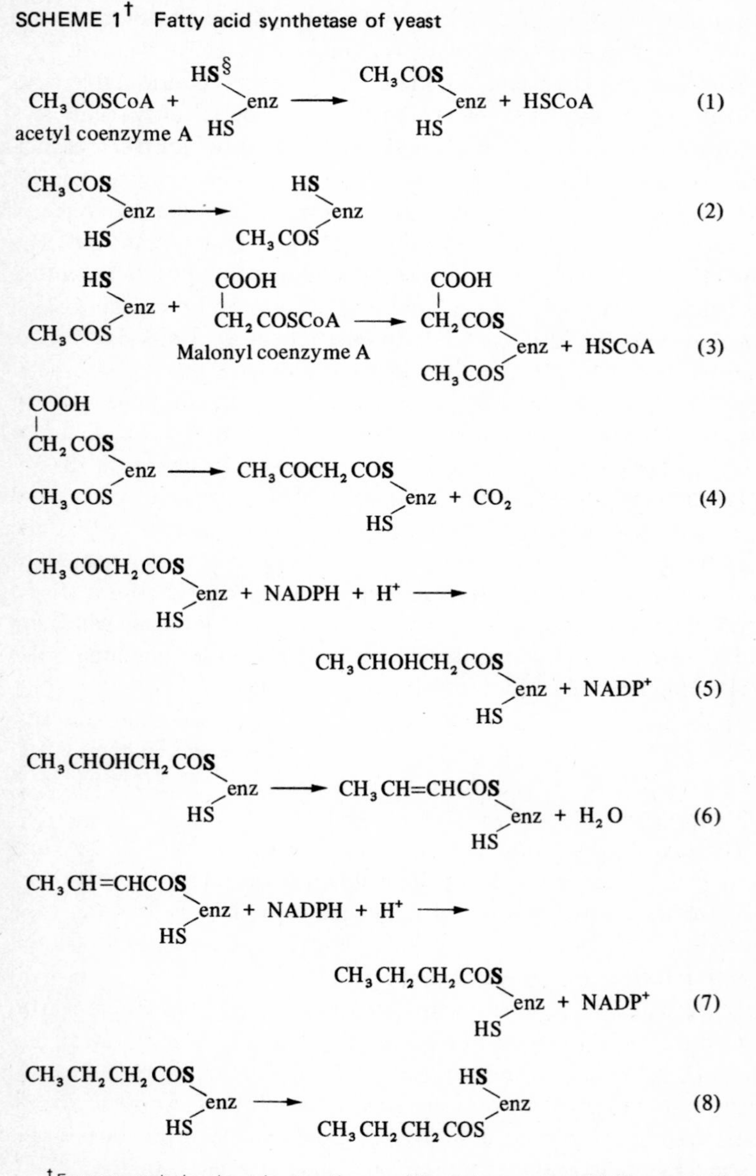

† Enzymes catalyzing the various reactions are (1) acetyl transacylase, (2) β-ketoacyl-ACP synthetase (ACP = acyl carrier protein isolated from *E. coli* systems), (3) malonyl transacylase, (4) β-ketoacyl-ACP synthetase, (5) β-ACP reductase, (6) β-hydroxyacyl-ACP dehydrase, (7) enoyl-ACP reductase; (8) β-ketoacyl-ACP synthetase.

§ The bold **S** indicates the "central" sulfhydryl group of the yeast enzyme complex.

fatty acids is similar to the biodegradation discussed in Sec. 5-1. However, different enzyme systems are involved.

In *E. coli* all enzymes involved in fatty acid synthesis are found in the soluble fraction of cell-free preparations, in contrast to the yeast multienzyme complex. As a consequence the strongest support for the above mechanism is derived from the mechanism of fatty acid synthesis in *E. coli* that has been definitely established.

In addition to the various enzymes for appropriate catalytic functions, acyl carrier protein (ACP-SH) plays a central role in *E. coli* fatty acid synthesis. The acyl carrier proteins from yeast are proteins of low molecular weight (16,000) and are characterized by the fact that a single 4′-phosphopantetheine residue (Fig. 6-6) is covalently bound in a phosphodiester linkage of a particular seryl hydroxyl group of the protein. The sulfhydryl group of this pantetheine residue is the functional group of the molecule, and it undergoes repeated esterification and deesterification during the synthesis of long-chain fatty acids. The overall reaction may be represented as

$$\text{AcetylCoA} + 7\text{malonylCoA} + 14\text{NADPH} + 14H^{+} \longrightarrow$$

$$\text{palmitate} + 14\text{NADP}^{+} + 8\text{CoA} + 7CO_2 + 6H_2O$$

The relationship between the acyl carrier protein of *E. coli* and the central sulfhydryl group of the yeast complex was recently established by showing that the central sulhydryl group is the sulhydryl group of a 4′-phosphopantetheine residue that is attached to an acyl carrier component of the multienzyme complex. Likewise there is a direct relationship between the sulfhydryl groups of *E. coli* β-ketoacyl-SACP and the peripheral sulfhydryl groups of the yeast fatty acid synthetase complex. Thus there seems to be a direct parallelism between the yeast multienzyme complex and the *E. coli* system where a complex is not known but where more information is available.

The fatty acid synthetase complexes from pigeon liver and rat liver have been obtained recently in essentially homogeneous states. The

$$^{-}O-\overset{\overset{O}{\|}}{\underset{\underset{O^{-}}{|}}{P}}-O-CH_2-\overset{\overset{CH_3}{|}}{\underset{\underset{CH_3}{|}}{C}}-\underset{\underset{OH}{|}}{CH}-\overset{\overset{O}{\|}}{C}-\overset{\overset{H}{|}}{N}-CH_2CH_2-\overset{\overset{O}{\|}}{C}-\overset{\overset{H}{|}}{N}-CH_2CH_2-SH$$

FIGURE 6-6 Structure of 4′-phosphopantetheine.

average molecular weights of these complexes are 450,000 and 540,000, respectively. They are therefore considerably smaller than the yeast synthetase complex. Both synthetases contain 1 mol of acyl carrier protein (4′-phosphopantetheine) per mole of enzyme, and they bind about two equivalents of acetyl and malonyl groups per mole of enzyme.

The importance of multienzyme systems, as opposed to single enzyme systems, stems from the ability of the former to catalyze a series of reactions, in sequence, for the biosynthesis of some material. Implicit in this argument is the assumption that the substrate, without moving, is in the proper stereochemistry for multiple reactions by different enzymatic activities. Many reactions require multiple steps for their completion. A simple enzyme can usually perform only one reaction. However, a multienzyme complex can catalyze many reactions in general and, if it is organized correctly, both stereochemically and chemically, can carry out multistep syntheses.

6-4 ARTIFICIAL ENZYMES

Enzymes may be synthesized from their constituent amino acids; for example, the enzyme ribonuclease which splits ribonucleic acid has been synthesized in two ways. One way which was accomplished by a group at Merck is a solution process, whereas the other method devised by Merrifield is one in which amino acids are added one at a time on a solid, polymeric support. There is a large effort in synthesizing enzymes from their constituent amino acids, but the subject at hand is a different one, namely, the synthesis, either chemical or enzymatic, of new enzymes that ordinarily do not exist in nature.

Probably the most worked on artificial enzyme is the enzyme thiolsubtilisin which has been formed from the bacterial proteinase subtilisin (which comes from *B. subtilis*) by a series of chemical transformations [Eqs. (6-5) to (6-7)] by two different research groups working independently. The reason that a bacterial proteinase was selected rather than a mammalian proteinase (e.g., chymotrypsin) is that the former contain no disulfide linkages, whereas the latter do, and one of the steps [Eq. (6-6)] involves a sulfide reagent which can lead to disulfide interchange and possible conformational changes in those enzymes which contain disulfide links.

The transformation of subtilisin to thiolsubtilisin involves the change of only one atom from oxygen to sulfur. It involves transformation of the amino acid serine to the amino acid cysteine. Thiolsubtilisin is not a very good enzyme: it is only about one-thirtieth as fast as subtilisin toward *p*-nitrophenyl ester substrates, and it shows no reactivity whatsoever toward amide substrates. It has been postulated that the larger size of the sulfur atom than of the oxygen atom leads to a slight change in stereochemistry which leads to these dramatic effects. Alternatively these effects could be explained by the difference in chemistry between sulfur and oxygen, for example, in hydrogen-bonding properties.

In order to transform a serine side chain into a cysteine side chain, the carbon-oxygen bond of the primary alcohol must be broken. The bond breaking can be facilitated by the activation of the hydroxyl group, for example, with a sulfonyl compound that yields a good leaving group [Eq. (6-5)]. This will inactivate the enzyme but will allow the subsequent replacement by a thiolate ion to occur [Eq. (6-6)]. The thiolate ion of choice is the thiolacetate ion that leads to the acetyl derivative of the thiol enzyme which spontaneously reacts with water to produce the free thiol enzyme [Eq. (6-7)]. These reactions are illustrated below.

$$\text{Subtilisin–OH} \xrightarrow[-\text{HF}]{+\,C_6H_5CH_2SO_2F} \text{subtilisin–OSO}_2\text{CH}_2\text{C}_6\text{H}_5 \qquad (6\text{-}5)$$

$$\text{Subtilisin–OSO}_2\text{CH}_2\text{C}_6\text{H}_5 \xrightarrow[-\,C_6H_5CH_2SO_3^-]{+\,CH_3\overset{\overset{\displaystyle O}{\|}}{C}S^-} \text{subtilisin–S}\overset{\overset{\displaystyle O}{\|}}{C}\text{CH}_3 \qquad (6\text{-}6)$$

$$\text{Subtilisin–S}\overset{\overset{\displaystyle O}{\|}}{C}\text{CH}_3 \xrightarrow[-\,CH_3CO_2^-]{+\,H_2O} \text{subtilisin–SH} \qquad (6\text{-}7)$$

There has been some controversy over whether thiolsubtilisin is, in fact, an enzyme for all substrates. However, there is no question that it speeds the cleavage of *p*-nitrophenyl ester substrates (faster than does cysteine), and so it must be a true enzyme.

Other bacterial proteinases have been used in the same transformation. They include subtilisin-Carlsberg, which comes from a different strain of bacteria, and the protease from *Aspergillus oryzae*. The same steps are again followed, and the resulting thiol enzymes show similar characteristics to those of thiolsubtilisin. This method thus seems to be a general one for the chemical transformation of an amino acid residue in enzymes. The reverse of this process, which is not known, would also be interesting.

The enzyme chymotrypsin has been modified in many ways. A methionine near the active site has been transformed into several derivatives, one of which is actually faster than the native enzyme. The histidine in the active site of chymotrypsin has been modified in several different ways. In the simplest modification a methyl group was introduced onto one of the nitrogen atoms of the histidine. This derivative, which may be called *N*-methyl-chymotrypsin, shows either no activity or one ten-thousandth of the activity of the native enzyme, according to two different people.

The enzyme trypsin has been transformed into an enzyme resembling chymotrypsin by chemically blocking the anionic specificity site of trypsin.

Many chemical modifications of enzymes are possible. Artificial enzymes can be made from natural enzymes. One enzyme can be transformed into another. Also, the possibility exists that a protein, which contains a binding site but no catalytic groups, may be transformed into an enzyme by the addition of appropriate catalytic groups in the correct stereochemistry.

In addition to chemical methods, an enzymatic procedure can be utilized in certain cases for replacing an amino acid residue. For example, the amino acid arginine was substituted for the amino acid lysine at the labile site of the soybean trypsin inhibitor. This molecule is of course not an enzyme, but rather the inhibitor of an enzyme, but it is a protein and has an active site, so it may be considered with the enzymes.

In the cleavage by trypsin of the soybean trypsin inhibitor, the arginine-isoleucine bond in the reactive site of the trypsin inhibitor is hydrolyzed and an active modified inhibitor results. This can be converted to an entirely inactive inhibitor (without the arginine) by the enzyme carboxypeptidase B. The driving force for the reverse reaction, which is known to occur, is the reaction of trypsin with resynthesized modified inhibitor to give a trypsin-trypsin inhibitor complex. Since lysine fulfills the specificity requirements of both the enzymes trypsin and carboxypeptidase B, it can replace arginine in the reaction. Thus, an enzymatic procedure can be used to produce a new protein.

The inhibitor from lima beans is double-headed in the sense that it naturally contains a sequence of amino acids in which an area containing the neutral amino acid leucine inhibits the enzyme chymotrypsin and another containing the cationic amino acid lysine inhibits the enzyme trypsin.

SUGGESTED READINGS

Bruice, T. C., and U. K. Pandit: Intramolecular Models Depicting the Kinetic Importance of "Fit" in Enzymatic Catalysis, *Proceedings of the National Academy of Sciences U.S.*, vol. 46, p. 402, 1960.

Knowles, J. R.: Intermolecular Forces and Enzyme Specificity, in "Proceedings of the Ninth European Peptide Symposium," E. Bricas (ed.), p. 310, North Holland Publishing Company, Amsterdam, 1968.

Jencks, W. P.: Strain and Conformation Change in Enzymatic Catalysis, in "Current Aspects of Biochemical Energetics," p. 273, N. O. Kaplan and E. P. Kennedy (eds.), Academic Press Inc., New York, 1966.

Gerhart, J. C., and H. K. Schachman: Distinct Subunits for the Regulation and Catalytic Activity of Aspartate Transcarbamylase, *Biochemistry*, vol. 4, p. 1054, 1964.

Koshland, D. E., Jr., G. Nemethy, and D. Filmer: Comparison of Experimental Binding Data and Theoretical Models in Proteins Containing Subunits, *Biochemistry*, vol. 5, p. 365, 1966.

Monod, J., J. P. Changeux, and F. Jacob: Allosteric Proteins and Cellular Control Systems, *Journal of Molecular Biology*, vol. 6, p. 306, 1963. These last three papers deal with the allosteric problem.

Changeux, J. P.: The Control of Biochemical Reactions, *Scientific American*, vol. 212, p. 36, 1965.

Ginsburg, A., and E. R. Stadtman: *Annual Review of Biochemistry*, vol. 39, p. 429, 1970. Multienzyme complexes, in particular the fatty acid synthetase complex, are discussed at a high level.

INDEX